Tourism and Australian Beach Cultures

TOURISM AND CULTURAL CHANGE

Series Editors: Professor Mike Robinson, *Institute of Archaeology and Antiquity, University of Birmingham, UK* and Dr Alison Phipps, *University of Glasgow, Scotland, UK*

Understanding tourism's relationships with culture(s) and vice versa, is of ever-increasing significance in a globalising world. This series will critically examine the dynamic inter-relationships between tourism and culture(s). Theoretical explorations, research-informed analyses, and detailed historical reviews from a variety of disciplinary perspectives are invited to consider such relationships.

Full details of all the books in this series and of all our other publications can be found on http://www.channelviewpublications.com, or by writing to Channel View Publications, St Nicholas House, 31–34 High Street, Bristol BS1 2AW, UK.

Tourism and Australian Beach Cultures

Revealing Bodies

Christine Metusela and Gordon Waitt

CHANNEL VIEW PUBLICATIONS
Bristol • Buffalo • Toronto

Library of Congress Cataloging in Publication Data
Metusela, Christine.
Tourism and Australian Beach cultures: Revealing Bodies/Christine Metusela and Gordon Waitt.
Tourism and Cultural Change: 30
Includes bibliographical references.
1. Tourism--Social aspects--Australia. 2. Beaches--Australia. 3. Human body--Social aspects--Australia.
I. Waitt, Gordon. II. Title.
G155.A75M47 2012
306.4'819099409146—dc23 2012009127

British Library Cataloguing in Publication Data
A catalogue entry for this book is available from the British Library.

ISBN-13: 978-1-84541-286-9 (hbk)
ISBN-13: 978-1-84541-285-2 (pbk)

Channel View Publications
UK: St Nicholas House, 31–34 High Street, Bristol BS1 2AW, UK.
USA: UTP, 2250 Military Road, Tonawanda, NY 14150, USA.
Canada: UTP, 5201 Dufferin Street, North York, Ontario M3H 5T8, Canada.

The policy of Multilingual Matters/Channel View Publications is to use papers that are natural, renewable and recyclable products, made from wood grown in sustainable forests. In the manufacturing process of our books, and to further support our policy, preference is given to printers that have FSC and PEFC Chain of Custody certification. The FSC and/or PEFC logos will appear on those books where full certification has been granted to the printer concerned.

Typeset by Techset Composition Ltd., Salisbury, UK.
Printed and bound in Great Britain by Short Run Press Ltd.

Contents

List of Figures

List of Tables

Acknowledgements

This book is the culmination of over a decade of research. We wrote this book with the assistance of many people. Thanks to the staff at the University of Wollongong Library, the National Library of Australia and the Wollongong City Council Library. Special thanks to staff in the Local Studies Area of Wollongong City Council Library who provided access to microfilm archives and local histories. And, our appreciation to Wollongong City Council Library, particularly Eva Westacott, for kindly giving permission to reproduce photographs from the Illawarra Images collection to illustrate our argument, and as art work for the cover. We are also indebted to those people who generated online library resources; specifically the digitised resources of the Australia Trove, National Library of Australia and the digital photographic collection, Illawarra Images, Wollongong City Council Library. Thanks to the support of the staff who maintain the archival records of Westpac and New South Wales State Rail. Thanks to the Illawarra Branch of the Surf Life Saving Association Australia for sharing their records. A special thanks to Christina Słŏ for giving us helpful feedback on earlier drafts and David Clifton for drawing the maps. We are also mindful of the behind-the-scene support of our families, who have helped us stay motivated.

Preface

This book explores the ever-changing relationships between bodies, oceans, beaches and tourism. The book contributes to the Tourism and Cultural Change series by taking a cultural geography approach to examine the emergence of sea bathing and sunbathing as a leisure and tourism activity. Our focus is the emergence of Australian beach cultures beyond metropolitan centres, where relatively little research has been conducted. We draw on a historical archive comprised of newspaper and magazine articles, bank records, rail records, council records, surf club records, tourist association records and guidebooks from 1830 to 1940 of the Illawarra, New South Wales, some 80 kilometres south of Sydney, to provide a snapshot of the shifting conceptions of beach tourism, bathing and the body.

Our contribution to the Tourism and Cultural Change series is twofold. First, drawing on the work of Elspeth Probyn we explore the reciprocal relationships between bodies and beaches; that is, how bodies help shape the beach, and, in turn, how the beach helps to shape bodies. The beach is not understood as a pre-existing entity, rather it is forged, maintained and challenged through an ever-changing constellation of uneven social relationships. At one level, our concern in exploring this reciprocal relationship is with the governance of nakedness in public. Among the British colonial gentry, revealing naked flesh in public, initially that of women and later that of men, was understood as disgusting, uncivilised and disrespectful. Drawing on the work of Michel Foucault, *Tourism and Australian Beach Cultures: Revealing Bodies* examines the various forms of governance over nakedness and semi-nudity – by colonial authorities, councils and later surf life-saving associations – to constitute bodies at the beach as respectable. At another level, our concern is with which bodies belong at the beach, and also those that are excluded. There are particular naturalised understandings of which bodies belong through how particular beaches become bounded and territorialised. Drawing on the ideas of Judith Butler, *Tourism and Australian Beach*

Cultures: Revealing Bodies offers a performative understanding of this territorialisation process by examining how bodies and the beach are co-constituted. Attention is given to examining the shifting intersection between aged, racialised, classed, sexed, gendered and national discourses that naturalise particular bodies as belonging on the beach.

Our second contribution is through our attention to the cultural economy of pleasure bathing. On the one hand, as in Britain, the railway is central to understanding the growth of seaside resorts on the margins of metropolitan centres in the 19th century. In New South Wales, the state-owned railway embraced the popularity of sea bathing to increase passenger flows on the South Coast Line, a line that was initially built to haul coal. The state-owned railway company played a key role in marketing the seaside resorts of the Illawarra, drawing on ideas of the picturesque and sublime to sell the pleasures of rail travel through the Illawarra as well as democratising travel by providing excursion fares and organising 'special' trains. Likewise, later, the automobile industry also sustained the Illawarra as a 'must see' tourism destination, particularly for day-trippers from Sydney, through publishing guidebooks and building lookouts.

On the other hand, council aldermen in the Illawarra did very little to encourage the growth of seaside resorts, ignoring the calls of various organisations and businesses lobbying for tourism from the early 1900s. The majority of aldermen framed surf bathing as a moral problem to be carefully governed rather than in economic terms. Drawing on modernisation and nation-building discourses, aldermen envisaged the economy in very narrow terms of the coal and steel industries. Regional economic plans and progress were fashioned by aldermen through the vision of becoming an industrial heartland of the Australian nation by increasing coal exports and manufacturing productivity rather than flows of tourists. Thought is given to the paradoxical qualities of Wollongong, the regional capital of the Illawarra, imagined as both the New Brighton of Australia and the Sheffield of the South.

Maps

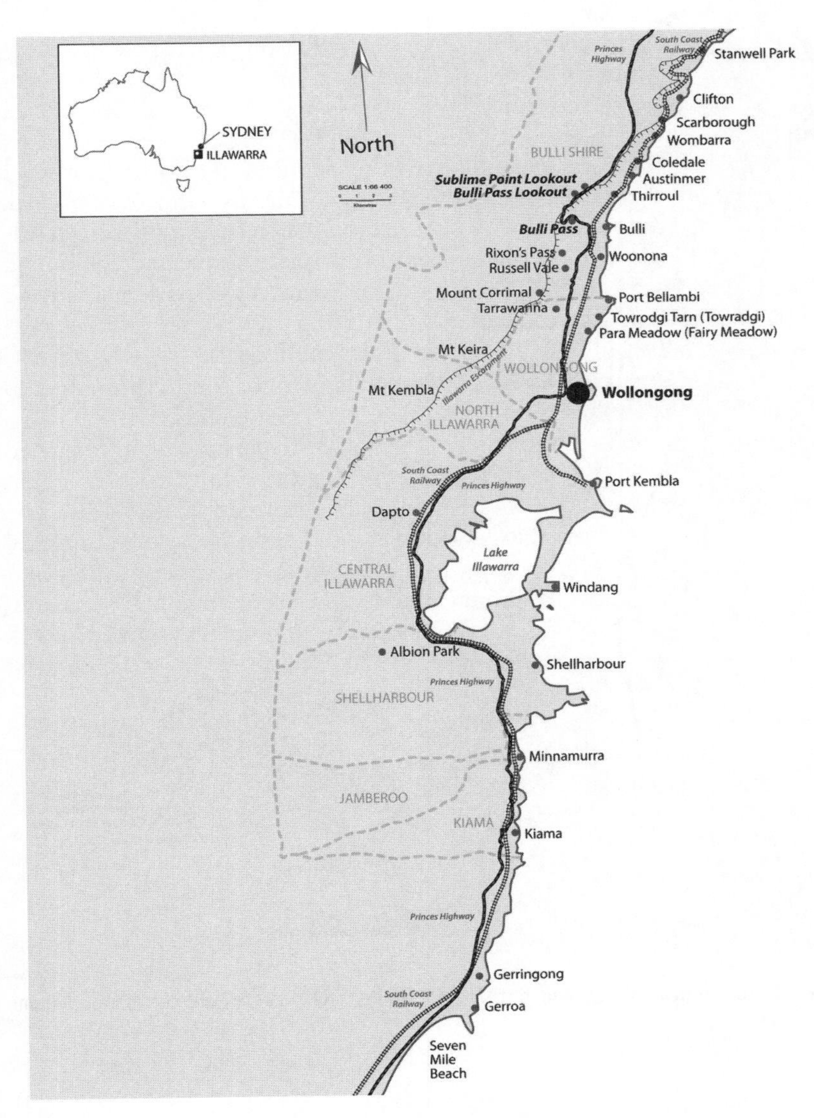

Map 1 Illawarra location map (*Source*: Mapped from archival record)

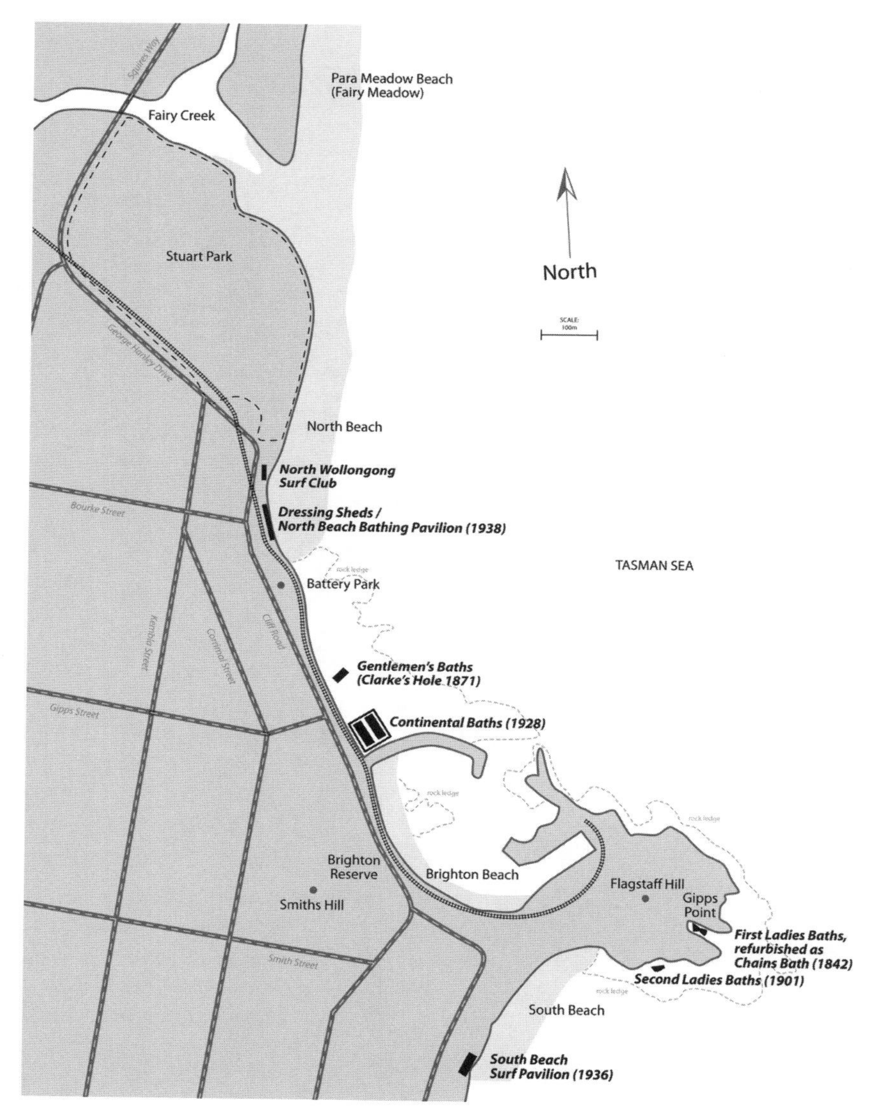

Map 2 Wollongong beaches and bathing facilities 1830–1940 (*Source*: Mapped from archival record)

Introduction: Stripping Off

This book explores the ever-changing interconnections between bodies, subjectivities, space, beach cultures and tourism. Drawing on the work of feminist scholars, our underlying premise is that space always informs, limits and produces the subjectivities of bodies (Probyn, 2003). That is, as bodies on the beach shape space, space is shaped by bodies on the beach. In this book we examine more closely how this occurs. Hence, this is a book that engages with the tussles over the geographies of the beach: its makings, boundaries and meanings for the West. Our focus is on how various people inhabited the beach through the emergence of beach cultures. Our context is Australian beach cultures beyond metropolitan centres from the early 19th century to the early 20th century. Out-of-the-way places and their people have long held sway in the popular imagination as tourism destinations. The beaches of the Illawarra on the east coast of Australia, some 80 kilometres south of Sydney, New South Wales, are no exception. The book critically interrogates how subjectivities of bodies are produced under very specific circumstances – the Illawarra beaches from 1830 to 1940. To do so requires identifying the ideological underpinnings of the very notion of the subject, and the ways in which bodies on the beach performed subjectivities.

The Illawarra beaches mark the edge of the Illawarra Coastal Plain, with the rocky outcrops and rainforests of the Illawarra Escarpment in the background (Bryant, 1981). The spatial focus of this book recognises that tourism practices in the Illawarra became fashioned by British and European geographical imaginations. For example, since the 1830s the Illawarra beaches were framed in the colonial imagination as the coastal idyll. Influenced by the Romantic Movement, the Illawarra was continually represented from the early 1800s in journals, paintings, guidebooks and tourist brochures as sublime, beautiful and picturesque. Reproducing European myths of the coastal idyll the Illawarra gained a reputation as a bathing resort in the colony of New South Wales. Since the early 1800s, the Illawarra had emerged as a preferred bathing resort destination among the Sydney middle class, and was dubbed by real-estate agents as 'the Brighton of Australia'. In the 1840s, Wollongong and Austinmer were framed by realtors as bathing resorts. However, that is not to imply that colonisation reproduced the British bathing resort in the colony of New South Wales. Instead, following the advice of Raymond Williams (1982), attention is given to the historical interplay of competing tendencies; the dominant, residual, oppositional and emergent. Taking Williams' advice, this book is about the aspirations, anxieties, fears and desires of bodies at the Antipodean beach.

The book draws on an historical archive from 1830 to 1940 of the Illawarra, New South Wales, to provide a snapshot of the controversies surrounding the touristic practices of the bathing resort; including bathing, tanning and lifesaving. This spatial and temporal focus is used to extrapolate the importance of multiple and intersecting economic, political, class, gender, sexuality and morality issues in fashioning the touristic practices of beach cultures in Australia in the late 1800s and early 1900s. Although the actions and practices reflect a discrete regional context, the spatial focus of this book offers an assessment of the cultural transmission of ideas and public debates about sea bathing and sunbathing not only across different geographical scales (bodies, region, nation and empire), but also in stabilising these scales.

At all Illawarra beaches today, swimming costumes, sunbathing, swimming and surfing are part of everyday life. Yet, from the early 1830s to around 1903, bathing bans prohibited anyone in the colony of New South Wales from legally stripping off their clothes and entering the ocean during daylight hours in public view between the hours of 6 am and 8 pm (legislated in the Acts of 4 William IV, No. 7 in 1833 and 2 Victoria II, No. 2 in 1838). At this time, drawing on privileged knowledge about gender, sexuality and respectability inspired by British bourgeois society, surf-bathing naked or semi-naked in *public* was considered by many of those in positions of colonial authority as immoral. However, paradoxically, in the early 1800s there were also bathing resorts – including those of Gerringong, Kiama and Wollongong in the Illawarra. Understandings of respectable middle-class masculinity and femininity were not troubled, so long as surf-bathing occurred in the privacy of an ocean bath, was segregated along the lines of sex and framed by a medical–moral alliance that understood purity in terms of hygiene rather than morality (Douglas, 1966). By the late 1800s, however, many young men allegedly bathed naked on public beaches. These men were reported in letters to newspapers as revelling in the opportunity to challenge understandings of respectability by revealing naked flesh in front of women visiting beach resorts. As Cameron White (2007) discusses, many middle-class men wrote letters on behalf of women who had been picnicking or promenading at beach resorts to express their disgust at the antics and bodies of bathers.

As a way of disciplining the naked bodies of bathers at the beach, in 1906 bathing ordinances were introduced and consequently the emergence of the public bathing reserve. Here, the surveillance of a beach inspector was to enforce a whole new set of rules regarding undressing and dressing the body at the beach. The flow of tourists to the Illawarra was sustained by the completion of the South Coast railway line that connected Sydney to Wollongong, in 1887, and Kiama, in 1888. The New South Wales Government Railways romanticised the Illawarra in tourism brochures and sold mass excursion rail travel to the working classes of Sydney to enjoy the pleasures of surf-bathing or bush-walking. At the turn of the 20th century, the automobile enabled touring Illawarra to be pitched as a series of sublime views for picnicking, or a 'must-see' attraction en route to Melbourne.

Between 1830 and 1940 the cultural, political, economic and social relationships forging concepts of the beach rapidly changed. This 110-year period is fascinating not only because of how surf-bathing was banned by bourgeois society as a way of expressing the civilising mission of colonialism, and then later regulated within the spatial boundaries of the public bathing reserve. Naked bodies bathing were understood as obscene by the civilised. Naked bodies at the beach were understood as a threat to the 'strength' and 'purity' of the British Empire. Fascination also arose from how certain styles of bathing bodies effectively reversed as British colonisers strove to define themselves as distinct from the empire, declaring the nation as the bearer of modernity. The new subjects of the surf-bather, surf-shooter, sun-basker and lifesaver all became central to justifying, legitimating, imagining and performing the Australian subject and nation (see Booth, 2001; Game, 1991; McGregor, 1994). What made possible these rapid transformations of bodies at the beach? The aim of *Tourism and Australian Beach Cultures: Revealing Bodies* is to explore this question.

In doing so this book contributes to recent scholarly work exploring how the beach became naturalised and bounded as an idyllic leisure place for British Australians, simultaneously marginalising Aboriginal Australians. Moreover, by focusing on an out-of-the-way place, this book seeks to address some of the silences in the writing and retelling of the beach in making and remaking of the imaginings of a white Australian nation. To date, the majority of beach texts provide a Sydney-centric narrative. Sydney beaches, such as Manly and Bondi have received most attention from scholars (see Booth, 2001; Dutton, 1985; Game, 1990, 1991; Huntsman, 2001; James, 1983; Wells, 1982; White, 2003, 2007). The Illawarra beaches are normally overlooked, with the important exception of Meredith Hutton (1997) and Louise McDermott (2005). Through a spatial focus on bodies at the beach, this book investigates the processes that enabled and constrained bathing and other leisure practices to unfold, or not, some 80 kilometres south of the metropolitan centre of Sydney, New South Wales, on the east coast of Australia.

The Illawarra beaches are also particularly interesting because of how they are entangled within underpinning ideologies of both colonial and national projects. Since the 1840s, bodies at the beaches in the Illawarra became the target of various projects of modernisation governed by initially colonial institutions, such as the Legislative Council, then later, national institutions such as the Australian Life Saving Association. Drawing on various ideologies, the regulations of various institutions were aimed at stimulating the governance of the self by setting out rules of conduct that sought to modify the appearance of naked and semi-naked bodies on public beaches. For example, drawing on Christian ethics, the men of colonial authorities argued that it was the 'purity' and 'strength' of middle-class bodies that were at stake. Following Michel Foucault (1978: 128) this was how the middle class sought to differentiate themselves from the working classes, 'not by the "sexual" quality of the body, but by the intensity of its repression'. Informed by the ideology of heterosexuality, nude mixed bathing of men and women was understood to undermine what was regarded as acceptable British sexual norms.

Following the passing of the 1894 Bathing Regulation Bill by the New South Wales Legislative Council, municipal authorities were given responsibility for enacting and policing State regulations at the beach. Six different municipal authorities regulated the Illawarra beaches through by-laws: Bulli Shire and Central Illawarra Shire and the municipalities of North Illawarra, Wollongong, Shellharbour, Jamberoo and Kiama. Hence, there was the possibility for variation depending upon how municipal authorities sought to manage the bodies of bathers. How bathing ordinances played out across the Illawarra and New South Wales was not a unitary project. Wollongong municipal authorities understood that semi-naked bodies of surf-bathers were immoral and lacking self-control. Even following Federation (1901), Wollongong Council sought to fix the boundaries defining middle-class bodies at the beach by forbidding sun-basking (sunbathing) and enforcing bathing costume regulations. Even when the colonial regime was being subverted by virtue of medical and eugenic discourses of the nation, the aldermen comprising Wollongong Council monopolised the moral high ground through declaring semi-naked bathing bodies unmanly and disgusting.

The English scientist Sir Francis Galton (1822–1911) first introduced the idea of eugenics in 1901 as a scientific means for achieving human betterment, even perfection. The term eugenics is closely aligned to the concept of race. Eugenics advocated that mental and physical improvement required maximising the inherited or distinctive qualities of a particular race. Family planning therefore was prescribed along racialised blood lines; that positioned people with 'mixed-bloods' as a weakening, or dissolving of the inborn qualities of the white race. Eugenics evoked a scientific authority by calling upon Charles Darwin's ideas of 'natural selection' and the rediscovery of Gregor Mendel's laws of heredity.

As discussed by Diana Wyndham (2003), eugenics was readily accepted in Australia and became particularly important in Australian population politics from the early 1900s to the 1930s. As illustrated by the White Australia Policy, the nation was forged by an unquestioned belief in the superiority of the white race. The eugenic promise of populating Australia with white Australians addressed a number of key political concerns following Federation – the diminishing birth rate amongst British colonisers, the unpopulated north, the fear of invasion from Asia and concerns of 'racial decay' spurred in part by environmental determinist arguments of degeneration of the white, colonising race under the tropical sun. This later fear was underpinned by belief that in a course of two or three generations the white race would become physically and mentally weaker because of the tropical environment (Soloway, 1995). Concerns of 'racial decay' also surrounded the increasing visibility of children of so-called 'mixed-racial' parentage, the 'half-caste'. The legacy of the stolen generation is testimony to the belief in the superiority of the white race amongst Federal and State authorities and the desire to populate Australia with a white population (Beresford & Omaji, 1998). Eugenics, therefore, not only built upon existing beliefs that left the superiority of the white race unchallenged, but provided ideas legitimated by science of how to boost the quantity and quality of the white population through racial segregation. The beach became forged by the surf lifesaving movement as one place that could assist the white race to achieve to their alleged 'natural advantages'.

Given that most aldermen in the Illawarra drew on Christian ideologies, this limited their understanding of surf-bathing bodies as obscene. Wollongong Council did little to sustain tourism infrastructure associated with bathing. Instead aldermen envisaged a regional economy and physical infrastructure tuned to specificities of coal and later manufacturing activities (Lee, 1997a, 1997b). Tourism advocates, such as the Wollongong Tourist Association and the South Coast Tourist Union had to develop arguments to defend resort tourism against a presumption of regional geography territorialised by coal and steel. From the late 1800s coal jetties were strung out along the northern Illawarra coastline. Further, from 1908, industrial activities intensified in Port

Kembla with the opening of the Electrolytic Refining and Smelting Company (ERS), followed by Metal Manufactures Limited in 1916 and Australian Iron and Steel (AIS) in 1927. In 1935, AIS merged with Broken Hill Associates (BHP), becoming the largest regional employee and source of revenue. Indeed, as Rob Castle (1997) explained, by the 1940s BHP and steel manufacturing underpinned explanations for why the Illawarra became one of the fastest-growing industrial regions in Australia. Therefore, colonial geographical imaginations of those promoting the touristic potential of the Illawarra as the 'New Brighton' faded. Taking dominance amongst key decision-makers was the discourse of the 'New Sheffield'; seeking to mirror the then international steel reputation of the city of Sheffield, South Yorkshire, England. The congestion, pollution and dangers associated with industrialisation while pitched in terms of 'progress', 'advance' and 'modernisation', worked against ideas of the more humble ontologies of holiday destinations that seemingly offered less speed, slower time, more fresh air, more sky and more 'nature'.

Nevertheless, the beach offered potential leisure opportunities and identities to the growing young population of Sydney and Wollongong employed in manufacturing activities; bathing/surfing masculinities and femininities. Despite the ways industrialisation restructured the Wollongong economy, the beach became an important site of longing and belonging in the lives of residents. Indeed, by 1940, the Illawarra Branch of the Surf Life Saving Association was the largest in New South Wales.

Our Approach

Despite the importance of understanding the rapid transformations of makings, boundaries and meanings that forge our ideas of the beach, historical analysis is almost completely neglected. Conceptually, this book picks up exciting feminist theoretical developments on bodies and processes of marginalisation in tourism studies and geography (see Browne, 2009; Johnston, 2001, 2005a; Waitt & Gorman-Murray, 2008; Waitt & Markwell, 2006; Waitt & Stapel, 2011). As Heidi Nast and Steve Pile (1998) noted, there is geopolitics to the body. Nast and Pile argued that the politics implicated in the metaphorical and material production of geographies at the scale of the body operates at two different levels; through the uneven and fractured process of connection and disconnection into places, and rights over the body (e.g. the gendered politics of body representation posed by questions of how to display the body through dressing, tanning or fitness regimes). There is a sexual, gendered, classed and racialised politics of the beach. Sexed, gendered, classed and racialised bodies are unevenly connected, and disconnected, through all spaces, including the beach. Particular attention is given in this book to highlighting the contested sexual politics of the beach, or what David Bell and Gill Valentine (1995: 1) term 'eroticised topographies'. The book addresses how individual bodies became included or excluded from the beach depending on how they were sexualised, gendered, racialised and classed. To do so, bodies and spaces of the beach are conceptualised as intimately tied together.

Spaces of the beach are conceptualised through Doreen Massey's (2005: 141) version of place as an 'event', a 'constellation of processes'. For Massey, place is far more than a location, or mapping the beach as a dot or line on a map. Instead, Massey encourages us to think about how places are thrown together – as a web of messy social interactions that have a spatial form. When places are conceptualised as a thrown together constellation or web of entangled social interactions, Massey presents us with the challenge of thinking about place both spatially and temporally. Following Massey, place is an outcome of negotiating not only the here and now, but is also interconnected with those 'far-off' times and spaces. Hence, drawing on Massey's ideas the beach becomes conceptualised as a cross-cutting and intersecting set of fluid, discursively produced connections, and disconnections, imbued with power that have a present and a past.

Bodies and bodily appearance is central to this book. As argued by Gill Valentine this is because the body:

> marks a *boundary* between the self and other, both in a literal physiological sense but also in a social sense. It is a personal *space*. A sensuous organ, the site of pleasure and pain around which social definitions of wellbeing, illness, happiness and health are constructed, it is our means for connecting with, and experiencing, other spaces. (Valentine, 2001: 15, italics in original)

The term 'body' is not understood in this book as a fixed biological entity. Instead, drawing on the work of Judith Butler (1993); Elizabeth Grosz (1994) and Elspeth Probyn (2000), feminist scholars have appreciated the impossibility to contain the materiality and meanings of bodies. Drawing on the work of Butler (1993), geographers such as David Bell, Jon Binnie, Julia Cream and Gill Valentine (1994); Clare Lewis and Steve Pile (1996); Robyn Longhurst (2001); Pamela Moss and Isobel Dyck (1996); and Gill Valentine (1996) noted the reciprocal connections between bodies and space; bodies shape space, as space simultaneously shapes bodies. Bodies are always bound up in space. The shaping and reshaping of bodies and space are conceptualised as mutually implicated with each other.

Furthermore, bodies are political because of how bodies become visible, take shape and are made sense of within the cultural practices of a particular time and space. Following Butler the 'materialities' of bodies as space and bodies in space, were conceptualised by these feminist geographers as performative, in the sense that their subjectivities are produced through the 'reiterative and citational practice by which discourse produces the effects that it names' (Butler 1993: 2). Bodies as space and bodies in space are therefore also bound up in questions of how the categories of sexuality, gender, class, age, ability and race are produced, reproduced and troubled through performative practices. Bodies as space and bodies in space are conceptualised as shaped and reshaped out of varied intersecting sexed, gendered, classed and racialised performances that may stabilise or destabilise these fluid categories. Critical tourism geographies of the body have begun to explore the way in which space and these categories are mutually constituted in and through events and destinations.

Our concern in this book is how bodies are made sense of and take shape within the cultural practice of bathing, tanning and lifesaving at the beach. In this book we open up some critical trajectories for examining the metaphorical and material tourism geographies of the body at the beach. The book discusses the politics and discourses within which bodies at the beach became intelligible. We explore the way in which bodies at the beach in the Illawarra from 1830 to 1940 were always bound up in questions of sexuality, gender, age and race. Bodies at the beach were always highly political. For example, bathing bodies in Sydney Cove in 1833 troubled the clarity of the boundaries between ontological categories of bourgeois society such as civilised/primitive, moral masculinity/immoral masculinity and dirtiness/cleanliness. The colonial authority intervened. Banning daylight bathing was intended to help hide the naked bathing body, understood as obscene and unmanly. Colonial bathing regulations can be conceived as a force of purification. Colonial geographies of the beach were therefore bound up in the enactment of geographies of ontological purity and boundary making that was diffracted in part through categories such as gender and race. In this book we are very much interested in the process of boundary making and territorialisation. Boundaries are not understood as fixed lines that divide a preconfigured us from them, civilised from primitive, urban from nature, culture from nature, spiritual from physical; rather, they are conceived as continually unfolding to help stabilise and secure the meaning and boundaries of the beach, and who is conceived of belonging. How the beach becomes culturally, socially and spatially bounded and territorialised is illustrative of how through particular cultural practices, bodies at the beach, are ascribed changing sexed, gendered, aged, racialised and classed identities. Bodies at the beach are therefore always political.

Our concern with the geopolitics of the body connects and draws upon the work of scholars who are positioned in a range of cognate disciplines but have focused on topics such as empire (Bashford, 2004; Blunt, 1994; Heath, 2010; McClintock, 1993; Michelson, 1993; Phillips, 2006; Ryan, 1997) and nationalism (Bonner *et al.*, 2001; Evers, 2008; Fiske *et al.*, 1987; Gibson, 2001; Hartley & Green, 2006; Kaplan, 1997; Morris, 1992, 1998; Perera, 2007, 2009) rather than tourism. One point of connection with these scholars is the work of Michel Foucault, particularly his ideas about governmentality. Foucault defined governmentality as:

> The ensemble formed by the institutions, procedures, analyses and reflections, the calculations and tactics that allow the exercise of this very specific albeit form of power, which has as its target the population, as its principal form of knowledge political economy, and as its essential technical means apparatuses of security. (Foucault, 1979: 20)

As Nicholas Rose (1996) pointed out, the notion of governmentality can be deployed to explore how self-surveillance is enmeshed within wider networks of power, such as institutions (the judiciary, the school, the family), discourses (medical, criminal, scientific, tourism) and analysis (surveys and statistics). Foucault regarded the main objective of governmentality as a form of surveillance and control aimed at stimulating the governance of the self through voluntary practices to maintain a healthy and productive population. As Foucault (1988) argued, the objective of governmentality is to produce compliant subjects that transform themselves in an 'improving direction', through providing them 'rational principles' that they should follow. For instance, how the colonial government sought to manage naked bathing bodies demonstrates how this colonial regime set about to 'purify' spaces of the beach by stabilising the boundaries between colonisers and colonised, and between moral and immoral masculinities of the colonisers.

Our concern with the geopolitics of the body also connects the opening up of tourism geographies to the importance of mobility. Governmentality is not entirely incompatible with John Urry's (2000) ideas of how bodies are continually formed and reformed within the institutions of the rail and road networks. The railway and automobile are significant for the Illawarra because they reconfigured time and space. Same-day travel became possible between Sydney and the South Coast, illustrating space–time compression. Furthermore, rail and car mobility involved distinct ways of dwelling, travelling and socialising. The railway and motor industries were central to formulating the subject of the tourist and the Illawarra as a tourism destination. We explore how the railway and motor industries engaged the ideology of

Romanticism to inform sets of travel practices that engaged people as tourists. This played out in establishing tourist circuits through guide-books and the opening of lookouts, walks and picnic sites.

Our Method

We rely on historical archives to pursue our inquiries. To tell the unfolding dramas in which the subjectivities on the Illawarra beaches are socially constituted the book draws on a range of historical archival materials including newspapers (e.g. *Illawarra Mercury, Kiama Independent* and the *South Coast Times*); *The Australian Women's Weekly* (first published on 10 June 1933); the New South Wales Government Railways records, the Bank of New South Wales records, Illawarra Surf Life Saving Association records, postcards/photographs and Council records. Amalgamation of the municipal authorities of the Illawarra meant that many of the Council records were destroyed. Many surf clubs lacked appropriate storage facilities. While the multiple sources comprising our archive allow the creation of a holistic picture of the beach, it is always partial.

Douglas Booth (2006: 101) called for 'a more cautious engagement with archived materials' and suggested that often archives are conceptualised as 'simple, straightforward sites of knowledge' (Booth, 2006: 92). Alert to such warnings, our analytical methods are linked to ideas associated with Foucault's ideas of discourse, knowledge and power. Hence, the archive is not conceived as a straightforward repository where truths can be retrieved. Instead, drawing on Foucault, discourse analysis enabled an interpretation of the archive. Following Foucault's ideas of discourse, knowledge and power, the archive provides insights to how particular understandings of the world may become taken as both legitimate and common sense. In other words, how particular persuasion operates to entail maintaining particular knowledge systems that operate to stabilise understandings of the world through making possible, or impossible, particular sets of social relationships. Foucault positioned the mutually interdependent relationship between power and knowledge as indistinguishable, arguing that: '[t]ruth isn't outside power ... Truth is a thing of this world: it is produced only by virtue of multiple forms of constraint. And it induces regular effects of power' (Foucault, 1980: 131). Hence questions about the 'truth' of knowledge are fruitless, for truth is unattainable. Instead, Foucault focused on questions addressing the effectiveness of sustaining knowledge (truth effects). According to Foucault, the mutual relationship between power and knowledge is underpinned by discursive structures.

Discursive structures are the relatively rule-bound sets of statements that impose limits on what gives meaning to concepts, objects, places, plants and animals (Phillips & Jørgensen, 2002). Foucault used these terms to refer to sets of ideas that typically inform dominant or commonsense understandings of, and interconnections between, people, places, plants, animals and things. Hence, while Foucault understood discourses to be inherently unstable, discursive structures are understood to 'fix' ideas of the world within particular social groups, at specific historical and spatial junctures. Furthermore, Foucault understood discourses as having power only when embedded within institutions:

> But this will to truth, like others systems of exclusion, relies on institutional support: it is both reinforced and accompanied by whole strata of practices such as pedagogy – naturally – the book-system, publishing, libraries, such as the learned societies in the past, and laboratories today. But it is probably even more profoundly accompanied by the manner in which knowledge is employed in a society, the way in which it is exploited, divided and, in some ways, attributed. (Foucault, 1972: 219)

For example, in the West, feminist scholars have demonstrated the importance of scientific thought in fashioning sets of ideas, or discursive structures, that naturalised dualistic thinking including men/women; heterosexual/homosexual, culture/nature and human/nature (Grosz, 1994; Rose, 1993). In sum, for both individuals and collectives, discursive structures establish limits, or operate as constraints, to the possible ways of being and becoming in the world by establishing normative meanings, attitudes and practices. Simply put, discursive structures are a subtle form of social power that fix, give apparent unity, constrain and/or naturalise as commonsense, particular ideas, attitudes and practices. Foucault referred to this form of social control as the 'effects of truth'.

Our approach allows us to provide a critical interpretation of how embodied practices performed by individuals at the beach are enmeshed in networks of power through an analysis of different 'texts'; bathing ordinances, newspaper articles, *The Australian Women's Weekly* articles, bank records, New South Wales Government Railways reports; surf lifesaving records as well as tourism and clothing industry advertisements. We illustrate the different and conflicting specific sets of ideas that come to define socially acceptable practices on the beach. The sets of ideas or discourses that fashion the mediascapes of newspapers and tourismscapes of the tourism industry, then, regularly construct and circulate representations to mobilise, normalise, configure, increase and stabilise particular ideas of who belongs at certain beaches in terms of race, sexuality, gender, pleasures and moral hierarchies.

It is important to acknowledge the implications of these texts for our analysis. First, the analysis produces partial, situated and interested knowledge. Crucially, the archive is comprised solely of official records kept by white, educated middle-class British-Australian men, often written for institutions that were integral to the colonisation of Australia. The archive is therefore classed, gendered and racialised. The activities of women are reported by men. There are also no insights gained from the working classes through the Great Depression years. Hence, the archive is comprised of a 'master' narrative that sustained a particular web of connection in which the beach was constituted. The voices of women and Aboriginal Australians are notably absent. There is no mention of the Indigenous people and their connection with the beach. As Grant Rodwell (1999) notes, British colonists blatantly disregarded Indigenous cultural attachments to the beach. For British colonists the beach was an empty page yet to be written on.

One way colonisation potentially continues to operate through the narratives contained in the archive is by silencing Aboriginal Australian activities and connections with the beach. As Denis Byrne and Maria Nugent argue:

> There is little exaggeration in saying that Aboriginal people are virtually invisible in the local post-contact landscape as described in archival records, in settler reminiscences, and in local histories. An illusion is created that they had vacated this landscape, leaving it as an open field for intensifying white occupation. (Byrne & Nugent, 2004: 11)

Hence, there is the danger of disregarding the Indigenous cultural and spiritual attachment to *Country*. Alert to the British colonial myth of *terra nullius* (nobody's land) it is therefore crucial to recognise that there were at least six Aboriginal tribes living in the Illawarra of the Dharawal language group: Dharawal, Wadi Wadi, Wandandian, Jerringa, Gurandada and Dharumba (Gibbs & Warne, 1995; Organ & Speechley, 1997). Further, it is vital to state that the whole of the Illawarra, and indeed Australia, morally belongs to Aboriginal Australians. Colleen McGloin (2005) has documented contemporary oral histories that outline how beaches are integral to Aboriginal stories and songs of the Dharawal Dreaming (*Alcheringa*). Recognising these limitations, the book then explores a host of themes through the conceptual lens of the beach as a constellation of represented social relationships: a stretching of social relationships, territorialisation, social boundaries, geographical imaginations, moral geographies, class, race and gender/sex differences.

Chapter Outline

Tourism and Australian Beach Cultures: Revealing Bodies explores the geographies of bodies in and through the spaces of the beach. Particular attention is given to how these geographies are mediated by cultural enactments of sexuality, gender, class and race. These themes are discussed in six chapters: (1) Sex in Private: 'Bathing in Perfection'; (2) The Public Bathing Reserve: Disciplining the 'Insatiable Desire to Pose on the Sands'; (3) Rail and Car Mobilities: Technologies of Movement and Touring the Sublime; (4) The 'Brighton of Australia' Becomes the 'Sheffield of the South': Knowledge, Power and the Production of an 'Industrial Heartland' in an 'Earthly Paradise'; (5) 'Battle for Honours': Surf Lifesavers, Masculinity, Performativity and Spatiality; and (6) Making Bathing 'Modern'.

The first chapter examines how the trajectories of tourism and beach cultures imported from bourgeois society in Britain and Europe played out in the colonial setting of the Illawarra in the 1800s. At this time, bathing bodies were situated within debates about obscenity within bourgeois society. Drawing on Butler's notion of 'performativity', attention is given to the normative expectation of how men and women would bathe at a resort in a specific way. Through the private spaces of the colonial bathing resort, men and women would demonstrate belonging to the colonial gentry through illustrating civility, dignity and modesty in themselves, and enforce it in others through regulating bathing practices in public. The bathing bodies of men who refused to comply with the daylight bathing bans and bathed naked at resorts caused gender trouble.

The second chapter draws on Foucault's (1978) arguments on 'regimes of truth', 'governmentality' and 'technologies of the self'. Bathing ordinances dating from 1906 that fashioned the beach as a bathing reserve are examined as an example of how sexed bodies were 'disciplined' to conform to the sexual norms of the Victorian bourgeoisie. In the first decade of the 20th century, according to the aldermen of the Illawarra Councils, only men wearing skirted-one-piece bathing costumes were welcome to bathe at public bathing reserves. Yet, at this time, most surf-bathers appeared to receive great pleasure from posing in front of crowds scantily dressed, rather than exercising self-surveillance along the lines of the social norms. The appeal for bathing amongst young people visiting the Illawarra appeared more to do with the beach being socially constructed as a highly sexualised space, rather than how surf-bodies were being refashioned on Sydney's beaches through the intersections of scientific, nationalistic and therapeutic discourses.

The third chapter draws on John Urry's (2000) ideas of mobility. Specifically, the chapter explores how rail and car mobilities reconfigured the geography of the Illawarra. The railway was crucial to facilitating mass tourism to the Illawarra. Excursion fares enabled mass participation in hiking, camping and surf-bathing. Guidebooks published by the New South Wales Government Railways used the familiar tropes of an earthly paradise to frame the rail journey through the Illawarra. Similarly, the automobile spawned a whole new tourism industry in the Illawarra. Touring the Illawarra by car in the early 20th century became a way to perform class distinction through picnicking and appreciating views as sublime.

The fourth chapter returns to Foucault's ideas of 'governmentality' and 'regime of truth' to explore the reluctance of Wollongong Council to support tourism activities as appropriate practices at the beach in the Illawarra in the early 20th century. This chapter explores how aldermen framed the Illawarra through economic modernisation discourses. Appropriate activities for the beach became coal loading and sewage disposal. Via the aldermen, a new 'regime of truth' emerged about the Illawarra as an Australian 'industrial heartland'. Wollongong framed as the 'Sheffield of the South' meant aldermen welcomed investment into the coal and steel industry, and resisted approving expenditure to support lifesaving on the beaches and actively discouraged the provision of facilities expected at seaside resorts in the early 1900s.

The fifth chapter returns to Butler's concept of performativity and the reciprocal relationships between space and subjectivity to explore the lifesaver as an effect of the weight of the sets of ideas, or regulatory fictions, of the Surf Life Saving Associations (SLSAs). The chapter explores why the surf beach became a privileged site of heteromasculinity and why it became naturalised as a masculine space. To do so the chapter outlines how the SLSAs drew on eugenic and militaristic discourses to establish and sustain normative expectations of who belongs at the surf beach as surf lifesaver, and why only men could do surf lifesaving. Surf lifesaving was transformed into a national project. An explanation is offered for why the semi-clad bodies of the surf lifesaver at the beach were never read as sexual or feminine, despite the shape and size of genitalia being outlined by the neck-to-knee regulatory bathing costume and rituals that involved displaying physical strength and skills. The final section investigates the contradictions surrounding the surf lifesaver at the Illawarra surf beach. Slippages are apparent in the surf records in terms of how bodies on Illawarra beaches did lifesaving. Evidence is presented on how hedonistic pleasures of tanning and surfing caused 'lifesaver trouble'. Surf lifesaving records reveal not only the continuities but also the discontinuities in the SLSA's project.

The sixth chapter examines the beach as a sexually contested space in the first decades of the 20th century. The relationship between the beach and sexuality is explored by examining different social groups attempt to make bathing 'modern'. The chapter is divided into five sections: bathing bodies, swimming bodies, fashionable bodies, swimsuit bodies and tanning bodies. Each section examines the discursive framings used to position Victorian bourgeois understandings of sexuality as repressive and outdated. Attention is given to the discursive framings of nature, science, fashion, cosmopolitanism, leisure and romantic love in constituting different heterosexual identities. For example, the swimming institution drew on science and discourses of physical culture and health to position the sports body as non-sexual; whereas the emergence of the transnational fashion industry gave greater importance to discourses of femininity and masculinity that disciplined body size, shape, skin colour and firmness. The beach is examined as a site where the boundaries between moral and immoral heterosexual identities were both ruptured and made resilient. An understanding of which bodies belonged at the beach is illustrated to be contradictory because of how sexuality intersects with age, fitness, race and age.

In sum, *Tourism and Australian Beach Cultures: Revealing Bodies* gives consideration to the continuities, discontinuities and transformations of Australian cultural practices at the beach. Adopting a spatial perspective has the potential to deepen understandings of the production and expression of identities in and through the beach. The book explores the body as a space and the body in space from a partial and situated perspective of an historical archive comprised for the Illawarra, New South Wales. Insights are provided into how daylight bathing in public was understood as outside the moral norms of colonial bourgeois society; how the bathing reserve was created in the late 19th century to prevent the sexualisation of bodies; how the Australian Surf Life Saving Association reshaped bodies at the beach through training programmes that conformed to a heteromasculinity configured by ancient Greek, scientific and national discourses as naturally Australian; and, how men dressed in trunks at the beach created a public outcry in the 1930s.

1 Sex in Private: 'Bathing in Perfection'

One autumn morning in 1895, Mr C. Holbrook, was charged for bathing within 'public view' of Barney and Manning Streets in Kiama, between 7 am and 8 am with 'insufficient dress' (*Kiama Independent*, 1895). Since 1833, in accordance to Act William IV, bathing during daylight in public view was banned in Sydney Cove and Darling Harbour by the New South Wales Legislative Assembly and extended in 1838 by Act Victoria II to all bathing resorts. In his defence Holbrook pointed to the lack of notification about the bans: 'No hours for bathing have been put up anywhere; no restrictions and no cautions; nor have any such appeared in any public newspaper or at any conspicuous place'. He went on to argue that: 'if the Act William IV should now be enforced ... then it will discourage visitors from coming to this town, many of whom come partly to bathe in the waves of the sea on beaches'. Holbrook pointed to the European therapeutic discourses to fashion the bodies of bathers, rather than bathing in public being understood as obscene. Mr Holbrook was determined to legalise daylight bathing from the beach, and redefine the signification of bathing bodies at colonial bathing resorts as the sight of the middle-class body at play. Holbrook was fined under the Towns Police Act (*Kiama Independent*, 1895). Daylight bathing became a site of contest between colonisers.

Since the opening of the first guest houses in the 1850s, the Port of Kiama, some 120 kilometres south of Sydney, was a popular bathing resort for affluent Sydneysiders who travelled by steamship to bathe in ocean baths. For Kiama, tourism in the 1850s was an important industry, alongside quarrying of blue metal and dairy farming (McDermott, 2005). Unlike the beaches of Sydney, the ocean baths of Kiama had not generated letters to the media from tourists or residents complaining of witnessing naked bodies as obscene. At Kiama, a sense of rightness and respectably was embedded through the designed layout of the ocean baths. During the late 19th century, allegations of the obscene were shielded against by keeping 'his' and 'her' bathing facilities not only in separate spaces in this bathing resort, but also hidden from public view. At Kiama, the act of bathing for the colonial gentry was shielded by a zone of privacy.

The aim of this chapter is to explore the daylight bathing ban. Why was bathing banned during daylight hours in the colony of New South Wales? How did bathing resorts respond to the ban on daylight bathing? Did the daylight bathing ban stop people from visiting bathing resorts? To answer these questions it is necessary to explore how bodies in public view at the beach were positioned through the interconnections between bathing, nakedness, sexuality and gender. At the most basic level the chapter examines nakedness. Rob Cover (2003: 55) argues that in the Judeo-Christian biblical tradition, nakedness 'was inseparable from sex and sexuality', and was 'located adjacent to the indecent, the obscene and the immoral'. In the Judeo-Christian biblical tradition nakedness is understood as the exposure of the genitals and cannot be separated from sexuality. As argued by Cover (2003), for nakedness to occur in the presence of a second party, it required establishing different sets of ideas or context within which nakedness could be shared for practical or pleasurable purposes. David Bell and Ruth Holliday (2000) outlined that in Britain until the mid-1800s, social nudity amongst men was the norm for bathers in many bathing resorts. The exposure to the sea was understood as 'therapeutic rather than pornographic' (Bell & Holliday, 2000: 132). In Australian colonial society, the naked bathing body on the beach became invested with classed, sexual and racial meanings. There was a spectrum of readings of nakedness that require registering.

Consequently, this chapter necessarily adopts a focus on empire. Ideas, fashions and technologies exported from Britain forged the establishment of resorts for the colonial gentry. Ocean baths were built informed by understandings of bathing as a therapeutic, pleasurable practice. As in England, bathing could occur at colonial bathing resorts amongst the gaze of others as a therapeutic practice without sexuality or shame. However, also exported from Britain were the ideologies, morals, values, habits, ideas and fashions that shaped the limits of race, gender and sexuality. This chapter explores how for the colonial gentry, bathing in public troubled the limits of race, class and gender.

The chapter draws on Judith Butler's understanding of gender as:

> the repeated stylization of the body, a set of repeated acts within a highly rigid regulatory frame that congeal over time to produce the appearance of substance, of a natural sort of being. (Butler, 1990: 33)

Butler argued that:

> acts, gestures, enactments, generally construed, are *performative* in the sense that the essence or identity that they otherwise purport to express are *fabrications* manufactured and sustained through corporeal signs and other discursive means. That the gendered body is performative suggests that it has no ontological status apart from the various acts which constitute its reality. This also suggests that if that reality is fabricated as an interior essence, that very interiority is an effect and function off a decidedly public and social discourse. (Butler, 1990: 136, italics in original)

Repeated performances of expected behaviours establish regulatory practices for bodies in the surf. Particular attention is given to how bathing bodies at the beach in the colony of New South Wales were rendered 'obscene' by bourgeois society. Bathing bodies at the beach were beyond accepted codes of public visibility. Bathing bodies subverted hegemonic sexed, gendered and classed construction of masculinity and caused 'bathing trouble'. This chapter explores the competing sets of ideas that defined bathing bodies at the beach. Turning attention to how gender is constituted through the performance of expected behaviours illustrates how gender is always unstable. From the 1830s, bathing bodies at the beach became a site where bourgeois conceptions of masculinity required constant securing.

The chapter is divided into two main sections. The first section examines the emergence and transformation of the bathing resort in Britain, specifically Brighton. Two transformations of the bathing resort are discussed, the medical beach and the pleasure beach. At the end of the 19th century, Brighton is argued to be an anomalous place where pleasure in part was derived from the playing with and challenging of the boundaries between moral and immoral sexualities.

The second section examines how the touristic desires, pleasures and practices of the English bathing resort were then exported through colonisation to the Illawarra; along with compulsory heterosexuality, race, social stratification, clock-time, bathing machines and rail. Ideas about the colonial bathing resort were deeply embedded in British society, but were transformed and distorted in the colonial context of New South Wales. Insights are given to how a peculiarly colonial version of the bathing resort arose in the Illawarra in the first decades of the 1800s because of how public displays of the body were constructed as obscene. Daylight bathing bans did not prevent the emergence of colonial bathing resorts. The bathing body of initially women, and later men, were simply hidden from public view in ocean baths behind fences. Bathing bodies were constituted as ostensibly non-sexual by virtue of excluding the gaze of the opposite sex through segregated bathing and the medical–moral gaze that shifted the focus of 'purity' from a moral question to a medical one. Bathing bodies in ocean baths that remained hidden from public view affirmed hegemonic constructions of gender for middle-class men and women. Men retained their dignity. Women retained their modesty. However, by the late 1800s, increasing number of bathers refused to comply with the daylight bathing ban. These bodies were defined in letters to the editors of the Illawarra and Sydney press as obscene because they broke the accepted codes of public visibility. Men in authority positioned the nude bodies of daylight bathers as examples of vice, social chaos, moral depravity and 'unmanly' behaviour.

Brighton, England

In England, investigating the touristic practice of sea bathing has already been cogently outlined by John Urry (1990) and John Walton (1983). Both Urry and Walton traced the touristic cultures of sea bathing that gave rise to bathing resorts to the intersection of sets of ideas about class, gender and health. The cultural history of the English beach resort was dependent upon the stratification of society through imposing class allegiances, sensibilities, values and tastes. They point to how the growth of the English bathing resort in the early 1800s enabled the aristocracy to remain distinguishable as social actors. The aristocratic clientele of popular English spa towns, such as Bath and Harrogate, were lured away to the sea after the spa became an appropriated model of middle-class self-hood. The spa town was one way respectable middle-class men and women defined themselves through appropriating classed sensibilities and touristic consumption practices: 'the spa habit spreading to the less polished of the gentry and the urban middle ranks' (Walton, 1983: 8). Dennis Jeans (1990) described how bathing resorts in early 19th century England, such as Brighton, Scarborough and Weymouth, not only replaced the spas such as Bath, as fashionable leisure destinations, but they also were shaped by an ubiquitous knowledge of class, gender, moral and health dynamics.

According to Jeans (1990: 277) the aristocracy followed royalty to the seaside 'beginning with the Prince Regent at Brighton'. In 1783, the prince was first taken to the coast to 'sample' the benefits of salt water and sea air. In the late 18th century the working beach of Brighthelmstone fishers was transformed into the medical beach of Brighton. The medical beach from which the gentry bathed became a temporarily domesticated space through the rituals of serving tea and picnicking. Bathing could occur among the medical gaze of others without sexuality by virtue of how bodies were framed and policed through carefully constructed codes of behaviour.

Drawing on Butler's (1990) concept of the embodied performative subject, the body is performed along the lines of highly stylised codes of behaviour which stabilise the semblance of subjecthood. Health practitioners produced one such way of performing and viewing the bathing subject. A whole genre of medical manuscripts advocated sea water. For example, Sir John Floyer, author of *History of Cold Bathing*, 'recommended the cold bath for nearly every malady in the medical dictionary' (Booth, 2001: 23). First published in 1702, Floyer's work reached its fifth edition by 1722 (Walton, 1983). According to Walton (1983: 11–12) a treatise written by Dr Richard Russell 'on the uses of sea-water as a near-panacea', took sea water 'from the status of an uncertain fringe medicine to that of a sovereign remedy'. In the context of bathing resorts, the medicalised discourse of the quality of sea water and sea air as a 'cure-all', as a match for any 'disorder' worked against the relationship between bathing subjects and bathing resort erotics. Instead, medical discourses framed how the bathing body was dressed, looked at, spoken about and publically made visible as a 'patient'.

From the late 1700s, bathing machines appeared as essential part of sea bathing etiquette at fashionable bathing resorts in England and elsewhere in Europe. The bathing machine was a wooden dressing-shed on wheels. Normally, horses were employed to roll the bathing machine into the ocean. In Europe, the bathing machine was a popular response to the circulation of narratives of feminine vulnerability and respectability. Women's access to public space was restricted by narratives that suggested 'the public' was an inherently dangerous place and reiterated the domestic as being safe. Hence, if women were to have access to feminine respectability, they either had to stay in the domestic realm, or be careful how they moved and appeared in public.

The ritualistic behaviours of the bathing machine worked to constrain the performative body so women could access feminine respectability. Working within a framework of heterosexuality, restriction of sexual arousal or activity was assumed to occur through gender segregation. Men and women were therefore normally segregated. Furthermore, disciplined by codes of respectable femininity, women bathed fully clothed to prevent even the barest glimpse of naked flesh.

The codes and rituals that inform the embodied performative subject of the bathing machine can be understood as efforts to ensure that ocean bathing, nor gazing upon bathing bodies were construed as sexual. The production of 'the respectable' was also bound up with restriction of mobility and the authorisation of the private spaces of bathing machine as legitimate spaces to bathe. Despite the codes and conventions that worked to separate bathing and bathing resorts from the sexual, there was always difficulty in preventing slippage between the sexual and non-sexual frames. Indeed, the unfolding popularity and pleasures of the bathing resort can in part be attributed to the instability of maintaining the discrete separation of the sexual and non-sexual.

By the mid-1850s the discursive framing of the medical beach was replaced by the pleasure beach (Shields, 1991). As Shields (1991); Urry (1990); Walton (1983); and Ward and Hardy (1986) all concurred, the regular visits of the Prince Regent (who became King George IV in 1820) to Brighthelmstone (Brighton) and the building of the Royal Pavilion to entertain Regency society, helped make visiting the beach both fashionable and pleasurable. As Rob Shields pointed out, the aristocrats fleeing the French Revolution bolstered the Prince's entourage, transforming 'the Brighthelmstone summers into a round of social events which rivalled the "Seasons" at Bath, not to mention those earlier in the century at Versailles' (Shields, 1991: 76). In the 1800s the gentry living in London visited Brighton as a weekend retreat. Pleasure resorts were about seeing and being seen. Many men and women travelled to bathing resorts to put themselves on display. Indeed, promenading gave rise in the 1800s to fashions only to be worn during a visit to a bathing resort (Haug, 2005). The way promenading was performed under the gaze of others operated to play with the boundaries of moral and immoral heterosexualities. As Debbie Ann Doyle argued, pleasure resorts of the mid- and late 1800s were 'complicated moral landscapes where people flirted with the line between playful sensuality and improper behaviour' (Doyle, 2005: 95).

Shields (1991) argued that Brighton as a pleasure resort became a place of sanctioned bodily excess, whereas Sally Munt (1995: 114) positioned Brighton as a site of sexual ambiguity, noting the 'camp, effeminate façades ... the orbicular tits of King George's Pavilion'. Both agree that the pleasure beach of Brighton in the late 1800s was a space where conventional moral regulations of the social order were temporarily suspended. The 1800s permissiveness facilitated the presence of pornography, prostitution, free love and homosexuality. Shields argued that accepted codes of public visibility could be challenged in Brighton in comparison with 'the more closely governed realms of the nation – the productive industrial areas, the "serious" world of London and the Parliament, or the "innocent" arcadian spaces of the agricultural counties' (Shields, 1991: 74). Consequently, he positioned Brighton as a marginal and liminal space, where cultural, social and spatial boundaries dissolve. Sue Morgan (2007) explored how the work of one evangelical religious association – The Ladies' Associations for the Care of Friendless Girls

(LACFGs) was to reconfigure the cultural boundaries of feminine respectability in Victorian Brighton. She discussed how LACFGs sought out 'immoral' behaviour in Brighton, offering lectures in chastity education. By drawing upon dominant Victorian ideologies of women as the spiritual and moral superior of men, the LACFG movement reappropriated Christian forms of respectable femininity bounded up in the reproduction of domestic space: virtue, pious motherhood, sanctity of the family, purity of the home, sacredness of marriage and loftiness of romantic heterosexual love.

Within the tourism literature, Brighton has also been understood as a carnival space, where the conventional social order is temporarily suspended, or inverted. Drawing on the ideas of Mikhail Bakhtin, Urry argued that:

> The fact that Brighton was the first resort in which the beach became constructed as a site for pleasure, for social mixing, for status reversals, for carnival, is one reason why in the first few decades of the twentieth century Brighton came to have a reputation for sexual excess and particularly for the 'dirty weekend'. (Urry, 1990: 31)

For Urry, the pleasure resort as carnival is understood as a site of social inversion. Possibilities exist for disrupting the moral boundaries of the obscene, the pleasurable and the desirable that regulated the divide between 'self' and 'other'. These qualities are still evident in the competing discourses of Brighton. In the cultural imaginary of Britain, Brighton is still imagined as the site of the 'dirty weekend', offering opportunities for heterosexual relationships deemed not to be legal, normal or healthy (Hemingway, 2006). Equally, Brighton is imagined as the gay capital of the South (Munt, 1995), subverting norms and public rituals associated with heterosexuality. The next section turns to how the practice of sea bathing played out in the colony of New South Wales. Alongside the export from Britain of the therapeutic practices of sea bathing that established colonial resorts as legitimate in the Illawarra were ideas about sexuality, class, gender and race that impinged on and differentiated bathing bodies in public as a problem.

The Illawarra Bathing Resort in the Australian Colony of New South Wales

In 1838, the New South Wales Legislative Assembly passed an Act (2 Victoria II, No. 2) prohibiting bathing 'within view of a public place or resort' between the hours of 6 am and 8 pm The prohibition on bathing in oceans 'within view of a public place or resort' (*New South Wales Government*

Gazette, 1838: 659) was an extension of an earlier ban in 1833 (4 William IV, No. 7) on the 'bathing in waters of Sydney Cove and Darling Harbour' in Sydney (*New South Wales Government Gazette*, 1833: 59), which at Federation was 'inscribed in Section 77 of the Police Offences Act, Act No. 5' (Booth, 1991: 157). Ocean daylight bathing was banned in the colony of New South Wales through the workings of colonial governance.

At one level, daylight bathing in public fractured the cultural boundaries demarking the coloniser and colonised. Ocean bathing became entangled within the grand moral premise that British colonists were bringing civilisation to a savage world. It was a social imperative, therefore, that colonists should at all times be seen as more 'civilised' than Aboriginal Australians. The spectre of the 'savage' that haunted the colonial imagination became flesh in the colonisers who relinquished their clothes and swam in public view during daylight hours. Clothing operated as a masquerade of civilisation. For those in authority, daylight bathing had to be banned to protect the race and empire.

At another level, surf-bathing practices mattered because clothes helped colonists define and redefine sexuality, gender and class distinction. Preoccupied with the construction of the properly socialised feminine and masculine body, bourgeois colonial society became obsessed with appearance. Styles of clothing were integral to maintaining social distinctions along the lines of gender, class, town and country (Russell, 2010). Drawing on Judeo-Christian bodily ethics, removing clothes and bathing nude or semi-attired in public was regarded as 'obscene' by the State. Following Victorian bourgeois conventions the visibility of genitals marked the body as sexual. Reflecting reproductive forms of moral heterosexuality, according to those in authority, sex belonged to the intimate relations of private personhood, allowing 'sex in public' to appear problematic. The daylight bathing ban involved remaking sex private by removing genitals from public display.

Furthermore, conventions of heteromasculinity blocked the possibility of bodies that unabashedly revelled in being sexy in public. As outlined by Penny Russell (2010), in Australian colonial society manners and dignity were integral to middle-class masculinity. Following etiquette from bourgeois British society, no respectable, rational and coherent man would expose their naked flesh to women and children. Therefore, men who bathed in public view were considered to be indecent because they were regarded as threatening the respectability and sensibilities of women. Equally, drawing on bourgeois English manners, men who watched women bathe were understood as 'unmanly'. At this time, as argued by Deana Heath working-class identities were understood in terms of 'a lack of self-reliance, ignorance, dishonesty, disloyalty or sexual impropriety' (Heath, 2010: 36). Moral regulation therefore operated not only through both formal legislation and policing but also informal ones. For example, through the work of shame as a mode of recognition of injustices committed against others, those who bathed in public during daylight may be exposed for failing to cover-up their bodies.

Cameron White (2007) wrote about the fears that arose amongst the colonial gentry regarding the social and moral implications of the nude public bathing practices of working-class bodies. These fears were heighted in the early 1800s because the body was understood as having porous boundaries. The body was understood as open to both physical and moral contagion (Bashford & Hooker, 2001). As Christopher Forth (2001) argued, at this time it was believed that moral contagion could be absorbed through corpuscles into the skin. Further, as Heath argued, the problem posed by moral contagion was one of boundaries – 'between the mind and the body, the inside and the outside, and the self and the other' (Heath, 2010: 36). The very embodied qualities of moral contagion troubled bourgeois masculinities and the crisis of boundary order that 'obscenities' created. Moral contagion was thus understood as unsettling the ability of individuals to self-regulate their own bodies, and to trouble the boundaries between culture and nature, civilised and wild.

Mobilised by fear, the aim of the bathing ban was to make the colonial beach 'pure' and 'civilised' by regulating nakedness which was framed as obscene and beyond the accepted codes of visibility. Operating through formal systems of governance (police officers) and informal ones (such as families), the daylight bathing ban is one example of moral regulation of colonial society as a spatial problem. Fear amongst the colonial bourgeoisie worked to contain bodies from bathing in the ocean during daylight hours. Following Michel Foucault (1991), the daylight bathing ban is one example of the British Empire as a biopolitical project, namely to regulate the obscene and stabilise the racialised and classed boundaries between particular moral and immoral heterosexual identities at the beach.

Archival records suggest that in 1838, the same year as bans on daylight bathing in public view were extended to resorts, Sydney estate agents were promoting the Illawarra, particularly Wollongong, as the Brighton of Australia with the aim of selling land. Advertisements of Sydney real estate agents framed the beach through medical discourse in the early 1800s. For example, in 1838, an article published in *The Australian* noted:

> The situation is beyond description healthy and delightful, and its proximity to the beach renders it peculiarly available for exercise, either pedestrian or equestrian ... JTW must not omit to point out the circumstance of its being in contemplation to establish this as one of the

> Watering places of this rising country; indeed Wollongong has so long been celebrated for its
>
> SEA BATHING
>
> that some of the wisest heads of New South Wales have years gone by foretold that it must eventually be the
>
> BRIGHTON OF AUSTRALIA
>
> (*The Australian*, 1838)

In 1841, again, *The Australian* reported that:

> This picturesque district (the Illawarra) will, it is hoped, become the resort of our Sydney fashionables and well deserve the appellation of the Brighton of N.S.W. [New South Wales] (*The Australian*, 1841, cited in Fleming, 1969: 5).

Throughout the 1830s and 1840s, beach-side blocks of land in the Illawarra were marketed to the middle classes of Sydney as summer holiday homes. As Cameron White (2003) discussed, a holiday home by the coast was a well-established trend amongst affluent Sydney families in the 1830s. From the 1830s, the real estate business paid particular attention to commodifying the Illawarra coastline as tranquil, 'closer to nature', and therapeutic.

Imagining Wollongong as the 'Brighton of Australia' continued to resonate in the 1840s with both officials and businesses. In 1840, Governor Bourke named the harbour foreshore 'Brighton Lawn Reserve', and the beach below became known as 'Brighton Beach'. Although Brighton never attained postal status in the Illawarra, businesses throughout the 19th century continued to imagine and sell the Illawarra as the 'Brighton of Australia' to help coax tourists. In 1856, hotel proprietor, Edward Johnson, renamed the existing beach-side hotel 'The Black Swan' as 'The Brighton' (Fleming, 1969). Through businesses imagining the Illawarra as 'The Brighton of Australia', the beach became entangled in the conventions and narratives of English bathing resorts. The beach was gazed upon through British colonial eyes as a site of pleasure, where social norms no longer applied. As discussed in Chapter 2 both the State and local Councils set about demarcating the unstable boundaries of the pleasure beach in an effort to 'purify' the moral appearance of bathers.

Middle-class Sydneysiders were also encouraged to visit the Illawarra as a beach resort, if not to invest in a second home, by following in the footsteps of Governor Lachlan Macquarie and colonial painters. As noted by Julia

Horne (2005), the Illawarra became a notable travel destination as a result of Lachlan Macquarie's Journal, titled *Journal of a Tour to the Cow Pastures and Illawarra in January 1822*. Macquarie's journal entry praised what he saw from Mount Keira as a 'very grand magnificent bird's eye view of the ocean' (Macquarie, 1956: 239). Macquarie's statement is literally and metaphorically dependent on a point of view. By adopting a 'bird's-eye view', he assumes a commanding imperial gaze over the Illawarra. As Richard Phillips noted, this bird's-eye perspective is a position beloved of British explorers because it naturalised the masculine authority of the coloniser, making the viewer 'monarch of all he surveys' (Phillips, 1997: 85). The viewer is empowered because of how they are positioned visually above and at the centre of things, yet apart from them. The viewer can assert their interpretation of the place from afar. From the 1820s, visitors to the Illawarra were made aware of the opportunity that the Illawarra Escarpment provided for a birds-eye perspective; an imperial and masculine gaze that imaginatively controlled and possessed the geography of the coast. As James Ryan (1997) argued, imperialism involved not only territorial acquisition, but also cultural formations, beliefs and practices. Imperialism in the Illawarra found sustenance in the tourist practice of the bird's-eye view of the coastline from the escarpment.

Colonial tourists could also draw on landscape paintings of colonial artists trained in the Romantic Movement in Europe to imagine and appreciate the Illawarra visually as 'beautiful', 'sublime' and 'picturesque'. As Brian Hudson (2000) noted, an appreciation of landscape with Romantic tastes was fashionable among British colonists. From the mid-1800s, the Illawarra became a fashionable destination for visitors with a Romantic appreciation of landscape given the relative remoteness from Sydney, the difficulty of access by land on horseback because of the escarpment and the presence of wooded gorges, a rocky escarpment, coastline, lakes and waterfalls.

Alan Menhennet (1981) argued, Romanticism is notoriously difficult to define. Very simply, Romanticism dates from the 1750s, and was an artistic, literary and intellectual movement. Romanticism provided theories of landscape aesthetic that valued often difficult to access locations, apparently lacking human impact, as 'beautiful', 'sublime' and 'picturesque'. According to Edmund Burke (1757), the characteristics of the 'beautiful' included delicacy, gradual variation, lightness, and smoothness, while the 'sublime' was associated with darkness, gloom, massiveness, ruggedness and vastness. The beautiful was built on pleasure, particularly the heterosexual desire of men. In contrast, the sublime was centred on pain and danger, to naturalise the conceptions of masculine difference. As pointed out by Simon Schama (1995) the sublime was associated with 'delightful horror' of self-preservation, the pleasure of the detached observer knowing that the potential for danger posed no immediate threat. As Horne (2005) discovered, the experiences of 'beauty' and 'sublime' throughout the 19th century were used to gender particular tourist activities based on the understanding that most upper-class women were motivated by a Romantic taste for delicate beauty in the scenery they sought, while men had a propensity to seek the sublime.

The picturesque was a concept introduced in English cultural debate in the late 18th century. The picturesque arises as mediator between the beauty/sublime binary. A number of artists, contributed to the debate surrounding the picturesque, including the Reverend William Gilpin and Sir Uvedale Price. The Reverend William Gilpin outlined rules of picturesque beauty and encouraged England's leisured travellers to seek picturesque scenes. Hudson provided a summary of William Gilpin's rules of the picturesque which included:

> a preference for rough textures, irregularity, partial concealment and the unexpected. Rocky, broken terrain was preferred to smooth land surfaces, natural woodland and weatherbeaten trees to carefully maintained lawns and groves. (Hudson, 2000: 75)

In addition, the picturesque could be enhanced by the presence of ruins, rustic bridges and quaint cottages. When artists trained in Romanticism arrived from Europe in the Illawarra, they had a set of rules by which to frame and arouse interest in both their work and the region.

As outlined by Orchard *et al.* (1994: 6) colonial painters from the 1820s portrayed the Illawarra as an earthly paradise – while the Illawarra was proclaimed by the governor of the colony, Richard Bourke, in 1834 as the 'Garden of New South Wales'. Amongst the artists who travelled south from Sydney were Augustus Earle (1827), Conrad Martens (1835, 1860) and Eugene von Guérard (1859). As argued by Orchard *et al.*, for artists working within the Romantic Movement 'the distinctive features of the escarpment and coastal landscape of Illawarra provoked many comparisons with picturesque European landscapes' (Orchard *et al.*, 1994: 5). The paintings of these artists undoubtedly encouraged visitors where to go looking in the Illawarra for the picturesque lagoons and view the mountainous sublime as illustrated through the opening of roads, tracks, footpaths and scenic lookouts. A Romantic interpretation of the Illawarra facilitated tourism flows. Informed by the Romantic landscape tradition from Europe, colonial painters to the Illawarra did not turn their attention to bathers. Their art did not transgress bourgeois standards nor attempt to subordinate moral and ethical standards to aesthetic ones (Figure 1.1).

Figure 1.1 Conrad Martens (1801–1878) 'Mt Keira' (1835). From the collections of the Wollongong City Library and the Illawarra Historical Society. Conrad Martens was one of many travelling artists who visited the Illawarra in the early 1800s. These artists were drawn to the Illawarra, understood through the European Romantic Movement as picturesque, who in turn circulated understandings of the Illawarra as picturesque through their paintings

Bathing Respectability

As elaborated by Lana Wells (1982), Walton (1983) and Jeans (1990), bathing was legitimised for the gentry as a medical and pleasurable practice, mimicking the ideas in England since the 18th century. Medicalised discourses of sea bathing were circulated in colonial newspapers. For example, the *Colonial Times* reported that 'many diseases, particularly of a cutaneous nature, have been removed by the frequency of salt-water ablutions' (*Colonial Times*, 1849, cited in Cumes, 1979: 298). Sea bathing as a therapeutic indulgence functioned as a mediator and metaphor of middle-class existence in colonial New South Wales.

Despite the daylight bathing ban, Wollongong, Kiama and Gerringong developed a reputation amongst colonial tourists as bathing resorts because of how bodies at ocean baths were not exposed to public scrutiny. As discussed by Marie-Louise McDermott (2005) and Jim Davidson and Peter Spearritt (2000), ocean baths play an integral part in Australian coastal tourism, offering safety from sharks, waves and rip-currents. At ocean baths the disciplining and surveillance [both self-surveillance and surveillance conducted by others; see Foucault, 1977] of bathing bodies functioned in ways that would not cause individual shame, collective moral offence nor 'gender trouble'. Bathing facilities were selected, built or upgraded with a *cordon sanitaire* around them to regulate the obscene by making the sexual body private.

As in Britain, compelled by the narrative of ideal heterosexuality bathing was segregated by sex. 'Mixed-bathing' undermined the patriarchal gender boundaries. In colonial New South Wales, the daylight bathing ban in public view reinforced 'appropriate' forms of femininity being bound up with narratives of feminine vulnerability. For colonial women to have access to feminine respectability, how they appeared in the 'public' spaces of the ocean bath was carefully regulated. The male voyeur was positioned as 'fearsome' and 'unmanly', threatening the sensibilities of women. Even without the daylight ban on bathing, the production of the threatening male voyeur and respectable forms of femininity worked to curtail the movement and visibility of bodies of women at the beach. In response to narratives of the heterosexualisation of space through the threatening voyeur, ocean baths were first designated for women to secure femininity as a delimitation of movement in public.

For example, an article that appeared in the *Sydney Morning Herald* in January 1842 reported on the work of convict labour organised by Governor Sir George Gipps at Wollongong's first 'Ladies Bath', Flagstaff Hill (Figure 1.2). The rock-bath was located in a cove at the base of cliff, under Gipps Point and therefore out of public view. According to the 'Illawarra Diary of Lady Jane Franklin', colonial gentry women and men had bathed here since the 1830s (Organ, 1988).

The author of the *Sydney Morning Herald* article with the initials G.U.A. discussed why the rock-bath was a legitimate space for women to bathe. Particular attention was given to discussing the refurbishments: the building of three bathing rooms, the upgrading of the path, installation of chains and ropes, construction of a 'rock curtain' around the outer edge of the pool and the building of a fence at the top of the cliff to screen bathers from public view. These measures and the designation of respectable femininity to the confines of the rock-bath can be understood as a response to the threatening encounters in public:

> delicacy, comfort and safety, to the most fastidious and timid have been attended to. A palisade fence was erected around the brow of the hill to intercept the view of the bathers from above. (*Sydney Morning Herald*, 1842)

Figure 1.2 Ladies' Bath (Chain Baths) Gipps Point, Flagstaff Hill, Wollongong (*Source*: Broadhurst Photographs. From the collections of the Wollongong City Library and the Illawarra Historical Society). Chains were installed across the baths with rings for bathers to hold. The chain and ring system were to prevent bathers from being washed out of the pool at high-tide

This article reproduced narratives of feminine vulnerability. The enclosed space of the rock-bath described as in terms of 'delicacy, comfort and safety' reproduced the anticipated injury to women in public spaces. The author understood that the rock-bath was 'admirably chosen', because privacy in public was seemingly assured. The article went on to declare that the 'ladies bath' was 'most perfect in the completion'. Having explained that bathing at this rock-bath posed no threat to the social ordering of society, the author went on to explain the benefits of visiting this out-of-the way place:

> The existence of such a bath, in such a district, so short a distance from Sydney, so easily and speedily arrived at ... that sea bathing, with change of scene and object are conjointly necessary to re-invigorate exhibited nature ... bathing in perfection ... and where throughout the Colony can these be enjoyed in equal purity, in greater variety, and richer luxuriance, than in Illawarra! – every taste can be indulged, every fancy and every whim can be satisfied. (*Sydney Morning Herald*, 1842)

Wollongong's first ladies' bath at Gipps Point, under Flagstaff Hill (a second was completed in 1901, on the southern side of Flagstaff Hill) was endorsed in the media by emphasising the proximity of the Illawarra to Sydney. Furthermore, proximity was aligned with the narratives of the therapeutic qualities of sea bathing, class distinction and the Illawarra as a refuge away from the oppressive city spaces. Wollongong was portrayed as simultaneously one of the most indulgent and purest sea bathing locations for colonial tourists. Compelled by the narratives of ideal heterosexuality, gender segregation, 'discrete' locations, the erection of fences and the medical–moral alliance were key mechanisms by which bathing culture could be represented as a safe, sanitised, desexualised space; a space of 'purity'.

In the pages of the *Illawarra Mercury* the bodies to be feared at the Wollongong rock-bath in the 1840s were those who failed to reproduce the ideals of moral heterosexuality. Threats were presented by those who departed from the social norms of compulsory heterosexuality – defined as the accumulative effect of the repetition of the narrative of heterosexuality as the ideal coupling. Bodies that refused to exercise self-surveillance in their orientation towards the ladies' baths caused 'bathing trouble'. Regulative norms functioned in ways that men watching women bathe defied decency. For example, in 1858, an anonymous author in the *Illawarra Mercury* reported that:

> During the whole of this summer, the ladies of the town have availed themselves very generally to the facilities afforded under the Flagstaff Point for enjoying the inestimable luxury of sea-bathing. Generally speaking, they are permitted to do so without being watched. The other day, however, we were informed that a well-dressed blackguard [blaggard] was seen prying and prowling about the bathing place whilst a lady was performing her ablutions. He was observed by the lady and caused her considerable fright and annoyance. Now this is a nuisance that ought to be stopped at once, for, although we only mention this case, still we are sorry to say that it is not of unfrequent (sic) occurrence. Such conduct is unmanly and despicable, and ought to be severely punished. (*Illawarra Mercury*, 1858)

A threat to the social order produced a transgression of the social norms of the ladies' bathing place. The anonymous reporter questions the masculinity of the 'prowler' for his failure to live up to the social norms of heterosexual masculinity. Understood as a sexualised act, moral heterosexual men did not gaze upon women bathing. Fuelled by the narrative of the prowler, women's bathing must be kept a private matter and thus remain pure.

By the late 1850s, the narratives questioning the respectability of bathing in public increasingly required that middle-class men at bathing resorts, as well as women, become increasingly discreet about their bathing bodies when viewed in public. To have access to masculine respectability, men also became increasingly restricted to swimming in ocean baths built out of public view. In Wollongong, in 1857, the introduction of a particular version of the bathing machine that appeared on Brighton Beach, Wollongong, gave men and women access to bourgeois feminine and masculinity respectability by keeping bodies private.

The bathing machine was operated by Edward Johnson, proprietor of the Brighton Hotel. His adaptation of the bathing machine, *'The Mermaid'*, was in essence a converted boat, described as a 'shed on wheels'. *'The Mermaid'* was '10 feet by 5 feet 6 inches', and was fitted with seats (Cousins, 1994: 196). On 29 December, 1856, Edward Johnson announced the pending launch of his bathing machine in the *Illawarra Mercury*:

> 'I'm Going To Bathe!' 'Where?'
>
> 'In Johnson's New Bathing Machine,
>
> The First Ever Seen in the Colony!!'
>
> The proprietor of the Brighton Hotel has the greatest pleasure in announcing to the inhabitants of, and visitors to, Wollongong, that he has constructed A Machine for Bathing, similar to those used at most of the fashionable Watering Places in England, which he confidently hopes will afford to all an opportunity of enjoying 'in strict Privacy and Security' the luxury of Sea Bathing at any hour of the day, and at any state of the Tide (*Illawarra Mercury*, 1856). The bathing machine was launched on the 5th of January 1857. It was dubbed 'the First Bathing Machine into New South Wales'. (*Illawarra Mercury*, 1857)

Paying to use a bathing machine enabled bathers to enjoy 'the luxury of Sea Bathing at any hour of the day'. Edward Johnson emphasised how the bathing machine brought 'strict Privacy and Security'. The bathing machine worked within the social norms that posited men and women to hide their bodies from public view during daylight. Appropriate forms of femininity and masculinity became bound up with the reproduction of private space as the only legitimate space for sea bathing. Purchasing a 'dip' at 1/- a dip, was one way to avert shame and reproduce a moral heterosexual subjectivity by keeping bathing a private act. Working to contain bodies in private was the fear of the bathing body being constituted by onlookers as uncivilised, savage, sexual or immoral. Bathers who confined their bodies to the private spaces of the bathing machine helped reproduce the boundaries of moral and immoral heterosexual subjectivities on the beach.

Yet, articles in the *Illawarra Mercury* point to some men refusing to confine their bodies to ocean baths, and causing 'bathing trouble'. From the 1880s, complaints became more numerous in newspapers about the threat to society of men who bathed in public view during daylight hours on Brighton Beach, Wollongong. For example, one anonymous letter to the editor of the *Illawarra Mercury* in 1888 remarked: 'the promenade on the top of the hill [Brighton Lawn Reserve] overlooking the sea' as being 'one of the best in the district'. The letter goes on to point out, 'many ladies *would* like to walk there or rest themselves on the seats placed close to the edge of the cliffs' (*Illawarra Mercury*, 1888, italics in original). However, the author complained that it was no longer possible to promenade here because they would witness the 'immoral practices' of men bathing. The anonymous author understood the bodies of these bathing men as breaking written and unwritten rules that limit what was acceptable bathing behaviour. Bathing in daylight defied decency. Exposing their bodies to women in public was understood as disrespectful. Men who bathed in public were expected to be modest. 'Gentlemen' bathers did not take pleasure in exposing their naked bodies to women. Indeed, according to an anonymous report in the *Illawarra Mercury* on 2 February 1869 in Wollongong there was a 'tacit understanding being arrived at that gentleman bathers shall bathe at a point to the north of the harbour' (*Illawarra Mercury*, 1869).

Such narratives called for action. As noted by McDermott (2005), such complaints in part explain the building of men's ocean baths in 'secluded' locations in Wollongong, Kiama and Gerringong. However, alongside efforts to spatially and visibly contain the sight of bathing bodies, the building of ocean baths also signalled the increasing popularity of swimming amongst the middle classes and the lobbying by Progress Associations. Hence, in 1871, the *Illawarra Mercury* reports of Mr Michael O'Toole's tender being accepted 'for the formation of the Gentlemen's Bathing Place on the north side of the harbour ... known as "Clarke's Hole"' (*Illawarra Mercury*, 1871). Following the building of the ocean baths, *Beautiful Illawarra: The Illawarra or South Coast Tourist Guide* described Wollongong as:

> fortunate in its bathing facilities. In a secluded spot at the foot of Flagstaff Hill there are both shallow and deep bathing places for children and ladies, with substantial sheds conveniently fitted up ... the gentlemen's bathing place, situated east of Smith's Hill. It has been excavated out of the solid rock, the depth varying from four to seven feet, and a concrete wall across the mouth enables the basin to remain full even at low tide. As the waves come rolling over the wall, the bather secures all the

> enjoyment without the danger of swimming in the open sea. (*Beautiful Illawarra*: *The Illawarra or South Coast Tourist Guide*, 1899: 68)

Similarly, in 1888, the Kiama Progress Association pressed for the construction of a men's ocean bath to promote tourism at Kiama, on Blowhole Point, 'in anticipation of flow of tourists' – from the completion of the Sydney to Kiama railway line (*Kiama Independent*, 1888). The Kiama Progress Association emphasised that:

> It certainly is a great disgrace to a town like Kiama to be without a good-sized sea bath, and now that the railway is open though to Sydney and a constant flow of visitors certain, we hope the matter will be taken in hand at once. (*Kiama Independent*, 1888)

On the one hand, the bodies of sea-bathers were bound up with the production of tourism space. On the other, the bodies of some sea-bathers were also understood as inherently dangerous by inhabiting public space during daylight hours. For example, in Kiama, shortly following the excavation of the ocean bath, some young men were convicted for 'bathing during prohibited hours in sight of a public place' (*Kiama Independent*, 1890). In response, the Bathing Committee set about planning to erect a fence at the Blow Hole Point baths 'so that the luxury of a sea bath may be enjoyed at any hour of the day' (*Kiama Independent*, 1890). However, the Crown Lands bailiff is reported to intimate that 'by erecting a fence, a trespass on Crown Lands would be committed' (*Kiama Independent*, 1890). No fence was erected.

Building sex-segregated ocean baths in 'secluded' places enabled bathing bodies to access genteel masculinity and femininity through making the sexual body private. Furthermore, bathing bodies in ocean baths increasingly became 'disciplined' by the sport of swimming. In the context of the ocean bath, gazing upon the semi-naked body at swimming carnivals was legitimised through the transformation of the swimmer's bodies into a mechanistic device through discourses of sport, health and medicine. At ocean baths, swimming as an emerging respectable middle-class sport provided frames in which a semi-naked body was legitimately gazed at by men and women without explicitly being linked with the sexual (Figure 1.3). As Susie Scott (2010) explained, the swimmer's body although scantily clad was concealed by a code of conduct. Following Norbert Elias (1994), etiquette at the swimming pool became understood as a facet of the civilising process. As Bell and Holliday (2000: 137) explained: 'the normalisation of nudity effectively desexualises it ... the body is re-naturalised and simultaneously de-sexualised'.

Figure 1.3 A swimming carnival at the Gentlemen's Bathing Place, Brighton Beach, Wollongong (c. 1910) (*Source*: Unknown. From the collections of the Wollongong City Library and the Illawarra Historical Society). In the distance, is Gipps Point, the location of the first ladies' bath in Wollongong

On the beach, however, nakedness among the gaze of others at this time was understood as illegitimate in the Illawarra. As discussed by White (2003) similar complaints were occurring at this time in newspapers about behaviour of naked bathers on Sydney beaches. Disgust was expressed in letters to editors at particularly men bathing naked during daylight hours and revelling in the interest of onlookers, particularly women. The disgust can be understood because the scantily clad bodies bathing were increasingly constituted as sexual. Men who took pleasure in exposing or flaunting their naked bodies on the beach caused 'bathing trouble'. In the context of colonial New South Wales in the late 1880s, the widespread appearance of the naked bathing body at the beach resulted in the sexualisation of the public sphere.

In Sydney, for example, in 1889, S. MacDonnell wrote to the *Sydney Morning Herald* to voice the opinions, he said, of 'the ladies of our boating community'. His letter outlined witnessing at Forty Baskets, Manly: 'hordes of naked persons who, on holidays and Sundays, take possession of our harbour beaches, rendering it impossible for any family party to land' (*Sydney Morning Herald*, 1889, cited in White, 2007: 24, 25). He went on to describe naked bathers as a 'disease' and a 'plague'. He concluded that the enforcement of the daylight bathing ban by the Mayor of Manly, by sending police officers to the beach 'would endear himself to the hearts of the gentler sex by his considerate action'. MacDonnell understood naked bathing bodies in daylight as immoral. And, as noted by White (2003), along with other middle-class men, MacDonnell argued that women, in particular, should be shielded from gazing upon the bodies of bathing men.

By the late 1890s, the behaviour of bathing men taking on public identities was a constant source of disapproval. As White (2003) argued, the masculinity of these men was often called into question through their portrayal in the media as 'larrikins'. The term larrikin was used in the 19th century to portray any immoral or irreligious mob in the colony. Larrikins were only intelligible as a group defined by its clothing, actions and semiotics against the bourgeoisie (Rickard, 1998). The pleasure that bathing men derived from exposing their bodies was an emblem of moral disorder in the public world. The regular appearance of the bodies of bathing men shifted the discourses surrounding bathing from the private to the public. Sex was a private matter and sexed bodies did not belong in the public realm. New legislation was required to cover-up the sexed bodies of men bathing in public. The corporeal boundaries that fabricated bathing bodies as sexual had to be disciplined.

In 1894, Mr Charles Heydon, lawyer and a member of the Legislative Council (1893–1900), introduced a Bathing Regulation Bill. Mr Heydon argued, that daylight bathing to occur without creating offence would require 'every person bathing [to] be clothed from the neck to the knee in a suitable bathing dress' (New South Wales Legislative Council, 1894: 1429). If surf-bathers were to take on a public identity then they must be deliberately dressed in ways that negated 'sexy' postures or sexual admiration. The neck-to-knee bathing costume was believed to be a mechanism by which the bodies of bathing men could fit within the prescribed sexual norms. In Mr Heydon's final address to the New South Wales Legislative Council on the 1894 Bathing Regulation Bill, he argued that failing to legislate a neck-to-knee bathing costume would allow 'the larrikins of Sydney' to continue to bathe with their 'abdomen bare and exposed to the view of females in the neighbourhood of public places' (New South Wales Legislative Council, 1894: 1429). Appearing clad in a neck-to-knee bathing costume in public was proposed as a way to restore a sense of order and decency to the colonial beach in New South Wales.

Conclusion

The colonial daylight bathing ban of 1838 was a way of disciplining sexual bodies at the beach. The daylight bathing ban affirmed the hegemonic construction of masculinity around manners and dignity. The beach became a place to regulate the obscene by 'keep[ing] gender in its place' (Butler, 1990: 34). At colonial bathing resorts of the Illawarra, the spatial layout of the bathing facilities assured that the performativity of bodies of men and women created a private space and respectable subjects. Newspaper reports in the 1840s were quite clear, the Illawarra resorts offered 'bathing in perfection'. In the private spaces of the resort the category 'sexual body' was affirmed; authorities segregated bathers by sex, and erected physical boundaries to prevent men from gazing upon the naked bodies of women bathers. The ban on daylight bathing in public view was an integral part of how the gentry regarded themselves as maintaining their class distinction and improving colonial society with their 'superior' values. Naked bodies on display in public were understood by bourgeois society as obscene, as well as a threat to the purity and social order of the imperial project.

At the same time, letters to editors attempted to limit the bathing bodies, particularly of men, who refused to comply with the daylight bathing ban. As these letters to editors suggest, these men acted inappropriately by showing off their naked bodies at the beach in public. Repeated requests were made by men, often on the behalf of women, for these bodies to be policed within the prescribed sexual and social norms of bourgeois society. These letters reiterated dominant discourses of masculinity by denouncing the act of bathing naked in public as 'unmanly'. Nakedness in public was the foundation of 'bathing trouble'. More specifically, the fabrication of male nudity in public being an obscene, unmanly body took on the status of truth.

Consequently, in 1894, Mr Heydon proposed to regulate the bodies of bathers in public by dressing them in a neck-to-knee costume. Clothes were to discourage pubic displays of bathing bodies as improper by covering the genitals. The next chapter explores the first decades of the 20th century and how bathing bodies were created and recreated through and by the spatial and regulatory practices of bathing ordinances. At the same time as the bathing body was being constituted through and by the regulations of the bathing reserve, different ways of doing bathing created gender trouble through declarations of the Australian Surf Life Saving Association (see Chapter 5). Tension was played out at the beach between colonial agendas of modernity and national desires. Following Federation in 1901, surf-bathing bodies as naturally Australian took on the status of truth in the wake of the desire to dissociate the Australian nation from the empire and justify its authority as the bearer of modernity.

2 The Public Bathing Reserve: Disciplining the 'Insatiable Desire to Pose on the Sands'

On Wednesday afternoon, South Beach, Wollongong was established, ceremonially as a bathing resort. A dressing shed which has been erected ... was declared ready for use. The Mayor (Alderman Lance) performed the ceremony. The modest little structure, he said, was a beginning and would soon be replaced by a building more pretentious
South Coast Times, 1914a

On a late spring day in 1914, the opening of a 'dressing shed' by Mayor Alderman Lance, at South Beach Wollongong, gave official sanction to the repositioning of the surf-bather as a welcomed visitor, so long as they obeyed a new set of bathing ordinances. The building of the dressing shed signalled attempts by the New South Wales Legislative Authority to 'discipline' surf-bathers. Many surf-bathers failed to exercise self-surveillance of the movement, posture and dressing of their bodies under the inspecting gaze of picnickers and promenaders at the beach. To accommodate the popularity of bathing at the beach within a bourgeois moral order, bathing ordinances created a designated space for surf-bodies. The only legitimate space for bathing in public view was the bathing reserve. Caroline Daley (2003) discussed the implications of a similar set of bathing regulations in New Zealand.

In New South Wales, the bathing reserve was in part a response to the failure of the daylight bathing ban. In the first decade of the 1900s, the delights of surf-bathing at Wollongong's beaches became very popular with young men and women from Sydney. The archival record suggests that young men and women disregarded social conventions of the day and were commonly bathing together on the beaches of Wollongong, following the practice at the turn of the century in French resorts such as Boulogne, Dieppe and Trouville. For the men and women employed in the growing industrial city of Sydney, the beaches of Wollongong were only a 51-mile train trip away. One tourist guide boasted that 'as a seaside resort, Wollongong undoubtedly offers attractions that cannot be surpassed' including 'basking in the sun on the silvery sea beaches' and 'splendid swimming baths for ladies and gentlemen' (*The Beautiful Illawarra District, Wollongong. The Ideal Seaside and Tourists' Resort*, 1910, no pagination). This popularity of bathing for pleasure is again reflected in the comments of Alderman White who is reported to have said: 'surf bathing has come to stay'; and Alderman Love who acknowledged that: '[S]urf bathing is going ahead all over the State' (*Kiama Independent*, 1908).

The bathing reserve can be conceived as part of a politically conservative project at work to address the increasing popularity of bathing for enjoyment in the surf. The bodies of surf-bathers became troublesome for those in authority because that which was commonly deemed to be private – the sexual body – was visible in public space. The bodies of surf-bathers dissolved any clear-cut distinction of public and private. Hence, the legislation of the bathing reserve sought to prescribe acceptable bodily postures, dress, movement and practices. Those individuals not willing to exercise self-surveillance and self-control would be disciplined by the beach inspector. The bathing reserve was illustrative of a form of territorialism characterised by claims of exclusivity by those in authority, through the assertions of social and moral superiority and distinction of genteel masculinity. The bathing reserve may be considered a defence of the moral order layered over the beach by the bourgeoisie in the grip of rapid social change brought by industrialisation and Federation. However, for some surf-bathers, the exclusivities of the bathing reserve sat uncomfortably with a new felt intensity of nationalistic, eugenic, therapeutic and scientific discourses of the beach. The beach was claimed by members of the lifesaving movement as 'home' for white Australians. The beach culture cult was mobilised to refashion surf-bodies within discourses of the new lifesaving movement as always-already Australian.

The concern of this chapter is to examine how the designation of the bathing reserve was used to govern surf-bathing for pleasure at bathing resorts in the first decades of the 20th century. In the first decade of the 20th century bathing bodies at the beach became a site where different conceptions of respectable femininity and masculinity required constant securing. This chapter explores the competing sets of sexed, gendered, classed and racialised ideas that defined bathing bodies at the beach. Through the legislation of the bathing reserve, surf-bodies were to be governed in a manner in which they became respectable to the ruling authority and limited the moral excess of surf-bathing. The chapter explores how bathing reserves sought to regulate surf-bathing bodies through by-laws that would discipline surf-bathers to exercise self-surveillance in public and conform to particular gendered norms.

The chapter draws on Michel Foucault's use of three terms; 'regime of truth', 'governmentality' and the 'technologies of the self'. Foucault (1980: 133) coined the term 'regimes of truth' to question the concept of universal truth. Instead, he understood 'truth' as produced, regulated, distributed and sustained by institutions invested with authority in a society. In Foucault's words:

> 'Truth' is linked in a circular relation with systems of power which produce and sustain it, and to effects of power which it induces and which extend it. A 'régime' of truth. (Foucault, 1980: 133)

'Regimes of truth' are important in this chapter because of how, first, British middle-class men held authority within the political and economic institutions that commanded the societal web at Federation. It was these men that held control of the distribution of 'truth'. Second, scientific discourses were increasingly becoming central in the creation of truth in western society, including Australia.

Foucault's (1991) notion of governmentality is underpinned by his working hypothesis of the mutual connection between forms of knowledge, institutions invested with authority, political decision making and power. At one level, his concept of governmentality points to the interconnectedness between how discursive structures are exercised to legitimise the authority of a government, to delineate 'problems' and the rationality to tackle the problem; including the procedures, institutions and legal forms. At another level, his concept of governmentality points to the mutual relationship between the sovereign statue and the autonomous individual. Following Foucault, government ranges from 'governing others' to 'governing the self'. Foucault's concept of governmentality provides a critical geopolitical angle for an analysis of the beach. First, the political rationality for governing the pleasures of the beach becomes the focus of study. In our case how a political rationality of Christian pastoral guidance in the early 20th century was challenged by the science of medicine. Second, our analysis must remain mindful to how distinction between the public domain of the State and the private sphere of society becomes configured and reconfigured on the beach. Third, attention is given to the strategies employed to govern society and render individual subjects 'responsible' (and also collectives, such as families and surf lifesaving associations).

Foucault's (1988) concept of 'technologies of the self' are devices, mechanical or otherwise of self-government. In Foucault's words 'technologies of the self', are devices:

> which permit individuals to effect by their own means or with the help of others a certain number of operations on their own bodies and souls, thoughts, conduct, and way of being, so as to transform themselves in order to attain a certain state of happiness, purity, wisdom, perfection, or immortality. (Foucault, 1988: 18)

Following Foucault, the mechanical and sets of ideas that constitute surf-bathing practices represents one such technology of the self. Surf bathing made possible the social construction of personal identities, most notably in this period the surf lifesaver. Technologies of the self underscores how a sense of personal identity is shaped through regularly adopting routines associated with going to the beach, including surfing, swimming or bathing. It is the level of commitment to these everyday routines or practices that is understood as integral to forging a personal identity. And, finally, these routines and practice facilitate pleasure through how they enable the possibility to care for oneself.

In addition, the chapter draws on Judith Butler's (1990) understanding of gender. Repeated performances of expected behaviours, established within the regulatory practices of the bathing reserve, work towards naturalising respectable bathing bodies on the beach. Particular attention is given to how some surf-bathing bodies disrupted notions of respectable masculinity and femininity. Those bathing bodies that subverted hegemonic gendered and classed constructions of masculinity and femininity caused 'bathing trouble'. In what follows, two questions are explored. First, what political rationality underpinned bathing regulations? Second, what strategies did the State employ to achieve a responsible and moral social order on the beach? For example, were the political goals of the State aligned with the emergence of surf clubs and the surf lifesaving movement?

Bathing Regulations

The 1894 Bathing Regulation Bill introduced to the New South Wales Legislative Council and subsequent amendments, aimed to bring a spatial order and coherence to the beach based on British bourgeoisie ideals of sexual restraint and modesty. Bathing legislation was a response to the failure of the daylight bathing bans to stop the practice of surf-bathing for pleasure and exhibitionism of semi-clad bathers (see Chapter 1). The social offence of surf-bathers rested on breaking the 19th century moral posture of bourgeois masculinity. The new bathing legislation relied on presenting bodies to each other in ways that attempted to define the socio-spatial context of the beach as non-sexual.

Following the 1894 Bathing Regulation Bill, councils became responsible for enacting and enforcing bathing legislation. The timing of the change of by-laws concerning daylight bathing was uneven across councils, often coinciding with the expiration of leases over 'baths'; that is any bathing enclosure or bath under council care and control. For example, Kiama Council's change of the by-law at the men's bath was reported in the *Kiama Reporter* in September 1905:

> Council resolves to rescind a by-law that prohibits bathing at men's baths between 10 a.m. and 4 p.m. (cited by Bayley, 1960, no pagination)

While from 1903 daylight bathing was legalised by councils, under the Bathing Regulation Bill *all* aspects of bathing were legislated, including: undressing and dressing, bathing attire, surf-bathing and sunbathing (sun-basking). As a way of attempting to bring social and moral order, bathing legislation had spatial implications. Underpinning the ways of organising spaces of the beach was the assumption of compulsory heterosexuality that coded the illusional order of sex/gender/desire and a moral order that aligned nakedness with the indecent. The spatial boundary of bathing reserves became highly regulated and the practices of (un)dressing and bathing segregated along the lines of gender and age.

Legislation seeking to regulate surf-bathing bodies, in a much deeper sense, was a way of taming the spatial. For the colonial bourgeoisie, stability and coherence over the beach frequently occurred at the intersection of three discourses; Romantic naturalism, primitivism and nakedness. The aesthetic and romantic appreciation of the beach, which in Australia emerged from Romanticism, and spilled over into popular bourgeois leisure pursuits like picnicking, effectively layered a moral topography over the beach (see Chapter 3). Romanticism rendered the beach as an escape from urban society by configuring it as pre-capitalist, drawing on the European fantasy of wilderness. As a backdrop for picnics the beach had strong romantic imaginings as 'wild' and 'natural', as pre-cultural and therefore untainted by civilisation. Imagined as pristine, nature became understood as an earthly paradise. The Romantic tradition of the picturesque found in the artistic and literary elites in colonial culture is detected in Governor Richard Bourke calling the Illawarra the 'Garden of New South Wales' in 1834 after an initial visit, in part because of the lush vegetation (Orchard *et al.*, 1994: 6).

Hand-in-hand with Romanticism was the environmental determinism of primitivism; where lush tropical vegetation was understood as offering a place to release one's sexuality and heightened sensuality. Consequently, these imaginings had already worked to eroticise bodies at the beach through discourses of primitivism. Constituted as a 'pre-cultural' site, the beach was imagined as 'free' from the moral constraints of colonial settler society, including bathing naked.

At Federation in Australia, biblical authority also still held sway on matters of nakedness amongst the bourgeoisie. Discourses of nakedness were framed by Judeo-Christian creation beliefs as found in *Genesis*. In the Garden of Eden, Adam and Eve were naked but innocent, named but 'not ashamed' (*Genesis* 2: 25). For Adam and Eve, shame about the naked self occurs only after Eve gave into her temptations and ate the apple, and then 'the eyes of them both were opened, and they knew that they were naked' (*Genesis* 3: 7). As Rob Cover (2003: 55) argued, following 'the biblical tradition – as read from our contemporary vantage-point', the shame surrounding nakedness is also understood as the exposure of the genitals in the presence of a gazing second party and 'cannot be disconnected from sexuality'. More specifically, nakedness in the biblical tradition was used to delineate the illegitimacy, and legitimacy, of certain expressions of sexuality. The Judeo-Christian bodily ethics that situated nakedness in terms of forbidden codes of sexuality was central to layering a moral geography over the beach at the time of Federation. Bathing ordinances were therefore also a form of sexual regulation that eased growing concerns amongst some middle-class men about the sexual agency of young women. As Douglas Booth (1997: 170) argued, 'Christian traditions located social order and stability in the renunciation and repression of hedonism'. Furthermore, as Bryan Turner (1984: 159) pointed out for many of the bourgeoisie at this time, civilisation was equated with asceticism and temperance that demanded the 'denial of the flesh and the control of emotion'. Covering the genitals was understood as essential to prevent moral decay.

Indeed, in the New South Wales Parliament Forum, sexual regulation had been increasingly debated since the 1880s, particularly the age of consent. In 1883, at the request of Sir Alfred Stephen, Arthur Renwick and William Charles Windeyer, the New South Wales parliament raised the age of consent for girls from 12 to 14 years (Allen, 1990). And, again in 1903, in the New South Wales Legislative Council, Sir Charles MacKellar tabled The Crimes (Girls' Protection) Amendment Bill proposing that the age of consent be raised from 14 to 17 years of age (Allen, 1990).

However, at this time, the age of consent parliamentary debate was framed through the ideology of sexism. Apparently, 'respectable gentlemen' would suffer from increasing the age of consent for women. The Attorney-General, B.R. Wise, claimed it was more urgent to protect young boys from seduction by 'vicious' girls. Following the biblical tradition, these young women were positioned as responsible for men's desire because they were unable to act rationally and control their sexual passions. The Attorney-General did not believe there were any large numbers of men going around seducing girls. He said that:

> the good sense and natural feeling of honour that there is amongst men not only restrains them as individuals, but the idea is repulsive and has a

> social stigma attached to it which prevents the commission of a crime of that kind. (Allen, 1990: 78)

According to Attorney-General Wise honour among men made it impossible for men to be responsible for underage sex. A redrafted Bill was eventually passed in 1910, that raised the age of consent to 16 years in line with Britain but it exempted charges being laid if the man believed the girl was 'immoral' or promiscuous and could be dismissed if the girl looked over 16 years of age. In the early 1900s, the prevailing concern in the courts over increasing the age of consent was that men would suffer. In the heteropatriarchal society of New South Wales, men governed the legality of sexual relationships. Furthermore, these debates over the age of consent provide insights to the assumptions made about women. Similar to Anne Summers' (1975) conclusions about Australian colonial society, women in the early 1900s were assumed to have only two roles: first, entrusted wives and mothers, who instilled Christian moral values and second, women as objects of sexual gratification.

Drawing on narratives of Romanticism, primitivism and nakedness, middle-class men in authority understood surf-bathers' bodies under the gaze of others at the beach as presenting potential threats to the social order, in-so-far as social encounters were understood as both erotic and sexual. Therefore, in order for people to bathe together in 'public space', the beach was regulated heavily to elude both nakedness and sexuality. To de-sexualise the beach, councils had a vested interest in the control over bathing reserves. By-laws were passed that necessitated the dividing-up and bounding of bathing reserves along the lines of sex, gender, age and attire. New enclosures were created including dressing sheds, sunbathing pens, and gender designated beach-ends. How people connected at the beach was negotiated through how the bourgeoisie came to understand forbidden codes of sexuality. The next section explores the by-laws of dressing sheds to illustrate the play of bourgeois moral codes in the social relationships which constituted bathing reserves.

Changing Clothes and Dressing Sheds: The Vanquishing of the Sexual

> A person shall not dress or undress or remove or disarrange any part of his bathing costume in any place open to the public view. (Clause 5, Ordinance No. 52, Local Government Act 1906)

Dressing sheds were one material expression of the bathing ordinances that layered a moral topography over the beach at the time of Federation in

Australia. For surf-bathers to maintain their 'respectability' under the gaze of others, Councils dictated where people could remove their clothes in a public, social setting. For the practical reason of changing clothes to bathe, the naked body was to be hidden away in dressing sheds. The first dressing sheds were designed to accommodate one person. Later, dressing enclosures or room became a legitimised site of nakedness. Here nakedness was shared for the practical purpose of showering and changing clothes. For nakedness to occur among the gaze of others without sexuality required establishing a space dissociated from the sexual (Figure 2.1).

Ordinance No. 52 (from the Local Government Act 1906, 'Public Baths and Bathing') worked within the ideology of heterosexuality and also actioned the idea that children over the age of eight were 'little adults' (e.g. see James Garbarino (1999) and Gareth Matthews (1994) for scholarship on children being treated as 'little adults'):

> The Council may mark off an area adjacent to any dressing enclosure or shed for the use of women, and may, by notice conspicuously exhibited thereon, forbid men above the age of 8 years to go upon any such area. (Clause 12 (d), Ordinance No. 52, Local Government Act 1906)

Figure 2.1 Dressing Shed, Brighton Beach, Wollongong (c. 1930) (*Source*: unknown. From the collections of the Wollongong City Library and the Illawarra Historical Society). In 1938 the dressing sheds were demolished and replaced with the North Beach Bathing Pavilion. Segregated by sex, the larger men's communal dressing and showering facilities were on the southern end of the building and were mostly open to the air

Men of honour aged eight years or over, for fear of upsetting the sensibilities of vulnerable women, would not undress in the women's dressing shed. The sight of an eight-year-old boy's penis was legislated as an inappropriate display of sexuality in the women's dressing shed. Hence, the dressing shed segregated along the lines of age and gender was one means to keep nakedness separate from the sexual at the bathing reserve. Dressing sheds were just one example of how Ordinance No. 52 'Public Baths and Bathing' sought to make bathing a respectable, civilised leisure activity.

'Nuisance inspectors' and dressing shed attendees were appointed to further insure that nakedness, nor the gaze upon it, was construed as sexual. Bathers who did not conform to these ideals were subject to criminal penalties. However, this is not to suggest that such a policing of rules to separate the nakedness from the sexual or the erotic was inevitably successful. The next section turns to bathing costume regulation.

Dressing-up for the Beach: A Spectacle of Masculinity or Encroachment of the Sexual

> All persons over 4 years of age bathing in any waters exposed to the public view . . . shall be clad in a bathing costume covering the body from the neck to the knee, so as to secure the observance of decency; and any inspector may require any person contravening this provision to resume at once his ordinary dress. (Clause 4a, Ordinance No. 52, Local Government Act 1906)

Bathing costume ordinances were particularly concerned with dressing men at the bathing reserve. Unlike middle-class men, Victorian social etiquette had always encouraged bourgeois women to bathe at resorts fully clothed to conceal the contours of their bodies. Bathing fully clothed was a necessity to prevent endangering their own heterosexual purity by arousing the sexual agency of men. Scantily clad young women supposedly presented a danger to society by encouraging masculine virility. In the late 1800s, under the watchful gaze of family, women wore bathing fashion imported from England that normally required wearing a three-piece suit at bathing places: 'a serge jacket, loose drawers that tied below the knee, with a skirt over the drawers' (Daley, 2003: 144). When the neck-to-knee bathing costume legislation was introduced, many women were already wearing far more clothing than legislation demanded at bathing reserves – including shoes, stockings and corsets.

Bathing dress regulations were informed by sets of values informed by Judeo-Christian genesis creation beliefs that necessitated an outward appearance of dignity and sexual restraint. From 1906, council by-laws specified regulations for neck-to-knee bathing costumes to be worn of materials that when wet would not become see-through by all bathers over the age of four years. In Foucauldian terms, bathing regulations were an example of a compulsion to discipline the naked body that looked in accord with middle-class discourses of respectability. According to a Christian infused temperance that required the denial of flesh, nakedness at the bathing reserve of anyone aged over four years of age was uncouth, impolite, sinful and a demise of civility.

For men, the one-piece 'bathing costumes', as they were known, were made of woollen and cotton fabrics and the full one-piece was designed to cover the body from neck to knee. Wet cotton and woollen fabrics used for early bathing costumes however could not hide the shape and size of the genitals of men. Consequently, how the genitals of men were revealed by the wet fabric of the body-hugging neck-to-knee costume become constituted as a problem. For many aldermen, bathing bodies of men clad in wet neck-to-knee costumes were sexualised bodies.

Dressing-up men in skirted tunics on the beaches of Sydney

The skirted bathing costume was welcomed by those who could not disconnect sexuality from how the wet-fabric of the neck-to-knee costume revealed the size and shape of male genitals. The skirt overcame the troublesome attributes of surf-bathers' bodies being gazed upon as sexual. Drawing on Victorian bourgeois ideas of sexuality, claims repeatedly occurred in the Sydney and Illawarra newspapers that the presentation of surf-bodies in public was improper. For example, G. Norton Russel criticised surf bathers for '[t]he paucity of covering worn, to the utter contempt and disgust of visitors to these seaside resorts' (*Sydney Morning Herald*, 1907a: 5).

Cameron White (2007) outlined how mayors of the beach-side Councils of Manly, Waverly and Randwick all initially seemingly supported the skirted-tunic for men, claiming the presentation of the semi-naked bodies of surf-bathers as indecent. For example, the Mayor of Waverly, R.G. Watkins, is reported to have said:

> After contact with water . . . the V-trunks favoured by many of the male bathers show up the figure . . . in a very much worse manner than if they were nude . . . people who patronise them should not be compelled to overlook bathers whom they do not agree with. (*Evening News*, 1907, cited by White, 2007: 26)

Watkins was horrified that women and children might see male genitals profiled by wet-costumes. For Watkins, it was seemingly impossible without a skirted-tunic to prevent the slippage of bathing; nakedness and gazing falling into the realm of the sexual and the immoral.

However, as discussed by White (2007), the proposal by three Sydney mayors to introduce skirted bathing costumes on Sydney's main beaches created great anxiety amongst some surf-bathers, particular those middle-class men who were members of the newly established Surf Bathing Association of New South Wales. Demonstrations opposing the skirt proposal were held by members of the Surf Bathing Association on 20 October, 1907, at Manly, Coogee and Bondi. These men swam in the surf rather than bathed as a form of recreation. These men questioned surf-bodies wearing a skirt in public, because it would mark them out as feminine in public. As discussed by White (2007) letters in support of the demonstrations were published in newspapers. An article in the *Daily Telegraph* quoted the voice of one surf-bather, Alfred Allen, who suggested that:

> Crowds would gather ... to watch the skirted or petticoated men-folk ... [in] the best ballet show yet presented to the public. (*Daily Telegraph*, 1907, cited by White, 2007: 30)

Similarly, a surf-bather writing a letter to the editor of the *Sydney Morning Herald* under the pseudonym, the 'Maroubra Marauder', opposed the idea of the skirted-bathing costume:

> Bathers are the most manly of men who love the exhilarating battle with the breakers ... They would not for a minute tolerate the wearing of women's clothes. The manly woman may be possible, but save us from the womanly man. (*Sydney Morning Herald*, 1907b: 5)

The Maroubra Marauder clearly understood the basis on which the skirted-costume was being legislated. His opposition was based on the reconfiguring of the ocean as dangerous and masculinity reconfigured in terms of the strong, virile warrior. Wearing a skirt therefore was understood to threaten the masculinity of bathers. The Maroubra Marauder's aversion to surf-bodies dressed in skirts could be understood in terms of his construction of masculinity. The skirted-surf bather challenged normative constructions of masculinity.

Members of the Surf Bathing Association drew on scientific, eugenic and nationalistic discourses to endorse surf-bathing, and surf-shooting (or body-surfing, that involves catching and riding a breaking wave without a surf-board) as a physical pursuit. In 1906, the *Sydney Mail* outlined:

> the irresistible fascination of 'shooting the breakers' – the moment of suspense before the rush of waters comes, the buoyant uplifting and tossing on the foaming crest, the restless drive up the shining beach until the sand is strewn with the panting bodies of bathers like blown leaves. (*Sydney Mail*, 1906a: 604)

Visitors to the beach were encouraged to gaze upon the shape, size, firmness and fitness of the bodies of surf-shooters, surf-bathers and surf lifesavers in terms of Australian standards of masculinity. As outlined by White (2007) the three Sydney mayors quickly backed down from introducing by-laws enforcing men to wear a skirted costume in the surf. This backpedalling is interpreted by White (2007) as an illustration of the changing significance of the exposed male body on the beach in terms of a desirable Australian masculinity.

The Sydney beach became a legitimate space for surf-bodies of men to be dressed in a one-piece bathing costume because of how the practice of surf-bathing produced flesh that was understood as masculine, healthy and desirable. For example, A.W. Relph, founder of the Manly Surf Club, and one of the most ardent advocates of surf-bathing as a sport, brought the surf-body into the customary social norms through the bushman myth. He, along with fellow Manly Surf Club member, W. Tonge, is reported to have stated that:

> But best of all ... what glorious dips we had at the ocean's edge, in the surf, the far-famed surf, and the fame of which will go on growing till at last Government and people alike will realise the value of this asset right here at our shores, its health-giving value, and its value in helping our youth to grow up fine, strong, hardy, shapely men and women ... an eminent member of the medical profession recently remarked, when lecturing to the Royal Society that 'no country in the world had such health resorts as Australia, and no country neglected them so much' ... Shooting the breakers is an art in itself. It is a sensation even more delightful than that experienced in a motor car or a toboggan slide. Surf-bathing is helping to build up a race of fine young hardy Australians, and everything should be done to encourage it. (*Sydney Morning Herald*, 1907c: 6)

Such statements helped to build a cult of the beach and symbolic national meaning of the racialised surf-body. Similarly, A.W. Relph wrote an article to the *Daily Telegraph* that looked to the beach as integral to the national landscape:

> Out back where there are no ocean beaches, where the call of the surf is not heard, the younger generation of Australia is well enough provided for ... But here in our city ... the outlet near at hand is the sea, and everything possible should be done by the authorities to encourage surf-bathing, not only for the sake of the health of the people, but also for the sake of the country, itself – for that country is greatest which has the most vigorous and healthy people. (*Daily Telegraph*, 1908: 5)

And a year later Relph again expressed how the nation should embrace the bodies of surf-bathers as a 'natural' part of the new Australian nation:

> for the pastime is not only a direct health restorer to the old and the young, but it is teaching the nation to love the freedom of the outdoor life, and [surf-bathing] ... is helping to build up a fine vigorous race from amongst the young people who live in the cities bordering on our shores. (*Sydney Morning Herald*, 1909: 5)

Relph's emphasis was on surf-bodies as national symbols. Consequently, looking at surf-bodies displaying their strength, skills and musculature could be naturalised as national pride. Surf-bodies were now interpellated at the intersection of discourses of health, race, physical fitness and Australianness.

Dressing-up men in skirted tunics on the beaches of the Illawarra

In Wollongong, the passing of the tunic bathing ordinance in Bondi and Coogee did not go unchallenged. One surf-bather expressed his concern surrounding skirted-costumes in an article published in the Illawarra Mercury. He also feared how men would be feminised by wearing a skirted-costume:

> Now the men on the other hand have held rigidly aloof from taking up women's pursuits as far as possible; much less have they attempted to imitate their dress. To be effeminate is their last desire. Imagine, therefore, the chagrin of the Coogee and Bondi men when they are themselves in a costume exactly similar to that worn by the ladies when they take their morning dip in the briny ocean. The thing is monstrous. Imagine a leading city professional man having to emerge from a dressing shed and make a beeline to the sheltering waves clad in a skirt. The

> awkwardness of it, too! How can he kick his legs about and perform aquatic feats before the admiring gaze of a 'mixed' assemblage with such an ungainly thing around his legs? But worse than all, the edict has gone forth from the adjoining municipalities that this has to be done. Ordinances have passed, and the thing is law. What is to be done if the Wollongong Council adopts a similar course? We do not like to contemplate such a calamity. (*Illawarra Mercury*, 1907)

For this anonymous surf-bather the skirted bathing-costume is positioned as a threat to the social distinction and respectability conferred to men by wearing suits and trousers. The display of the size and shape of male genitalia through wearing bathing costumes was understood as manly rather than immoral. The masculine virility of the surfing body was a social norm circulated in the *Illawarra Mercury*. No mention was made in the Illawarra print media of enforcing the new skirted bathing costume regulations that hid the penis and bottom. In January 1908, the *South Coast Times* simply highlighted that all bathers had to wear neck-to-knee costumes under Ordinance 52:

> From the Local Government Department there was a copy of ordinance for the regulation of bathing, as gazetted. The ordinance provides that bathing costume must cover the body from the neck to the knee. (*South Coast Times*, 1908a)

Another article that reiterated bathing ordinances a month later in the *Illawarra Mercury* similarly made no mention of skirted bathing costumes for men. Instead, the article reported that: 'The first rule is that all persons over eight years of age ... shall be clad in costume covering the body from neck to knee' (*Illawarra Mercury*, 1908a). The appearance of these articles suggests that failure of bathing ordinances informed the political rationality of biblical traditions to control populations socially. However, the guiding rationalities of the pleasures of surf bathing were also informed by the 'truths' of science, eugenics and nation. These alternative sets of ideas positioned the display of the fit, taught, trim, muscular body as manly. As argued by Booth (2001: 48) surf-bathers generally ignored regulations 'that were rarely policed and occasionally mocked'. Enforcing young men to conform to the norms of a one-piece costume was almost impossible, yet alone a skirted-bathing costume. In Foucauldian terms, the bodies of male surf-bathers refused to be 'disciplined' – and conform to the gendered norms from British colonial society.

Indeed, over a year later Alderman Lance lamented on the failure of surf-bathers to obey dress-regulations at Wollongong's bathing reserves. In March

1909, he is reported as finding gazing upon surf-bodies as 'very demoralizing' (*Illawarra Mercury*, 1909). He understood the bodies of surf-bathers as immoral because of the sight of male genitalia shapes and sizes. He went on to outline the implication for Stuart Park, North Wollongong Beach, of the pleasures seemingly derived by some surf-bathers displaying their bodies to the gaze of picnickers:

> That beautiful spot – Stuart Park – there was no use for now, for previous to the inauguration of surf-bathing the park was found to be thronged with young people taking a walk. (*Illawarra Mercury*, 1909)

For Alderman Lance the failure of surf-bathers to exercise self-surveillance and self-discipline according to the bathing ordinances made Stuart Park useless for picnics or promenading.

Bathed in Sex: The Gaze at the Surf-Bather as a Sexual Act

> A man above the age of eight years shall not trespass upon any part of a public bathing reserve which is set apart for women. (Clause 6 (a), Ordinance No. 52, Local Government Act 1906)

> A woman shall not trespass upon any part of a public bathing reserve which is set apart for men. (Clause 6 (b), Ordinance No. 52, Local Government Act 1906)

Following the ideas of Foucault (1977), bathing regulations can be understood to cultivate not only the attributes, movement and abilities of the bodies on display at the bathing reserve, but also how the body is looked upon. As a means to prevent the encroachment of the sexual and the erotic gaze, strict regulation and further policing was required of both surf-bathing and sunbathing in the reserve. As one means to keep the encounters at the bathing reserve non-sexual, working within the cultural system of gender dualism, bathing regulation required the separation of men and women. The policing against sexuality in the bathing reserve was entangled in the man/woman binary. As a means of keeping the dressing sheds non-sexual the separation of genders was essential. Clause 6 illustrated Butler's (1990) heterosexual matrix in which the illusionary order of compulsory heterosexuality codes sexuality. To prevent nakedness from being read by others as sexualised or eroticised, dressing rooms were spatially organised along strict age and gender lines.

Under Clause 6 of Ordinance No. 52, Local Government Act 1906, municipal authorities passed by-laws to legislate for the separation of bathers by sex. Lana Wells (1982) discussed how Waverly Municipal Council attempted to police against sexuality and the erotic gaze by creating exclusive bathing reserves for men and women at separate 'ends' of beaches. As discussed in Chapter 1, in Wollongong, men and women in the mid- to late 1800s had been encouraged to bathe in the sex-segregated municipal-owned baths. Similarly, from the late 1890s in Kiama, bathers were supported to follow the gendered moral topography layered over the ocean baths. Women bathed at Pheasant Point, while men bathed at Blow Hole Point (McDermott, 2005). In Foucault's terms, the sex-segregated bathing, alongside the neck-to-knee bathing costume regulations, were disciplinary mechanisms through which the desexualisation of communal bathing was achieved; rule-following enabled men and women to enjoy the pleasures of surf-bathing without their respectability being questioned and performed a disciplined gaze that was allegedly without interest in the sexual. As noted by Booth (2001) segregated bathing provided opportunities for women to become used to seeing themselves in a bathing costume and to learn how to swim away from the heterosexual gaze of men.

Nevertheless, the two-gender exclusiveness of the bathing reserve disclosed the social construction of the bathing reserve as a sexual site; the implicit eroticism encoded in the beach, the positioning of many young women as promiscuous, and the display of genitals as sexual. For some, surf-bathing and gazing upon surf-bathers, could not be understood as an exclusively non-sexual form of pleasurable activity. Without gender exclusions to exclude the presence of heterosexual desire, surf-bathing was charged with creating a social space for unsupervised flirtation and a sexual gaze.

Bathing ordinances were designed to prevent gazing upon surf-bathers slipping into a sexual frame. However, newspaper articles suggest that bathing by-laws were flouted by some men who refused to exercise self-control and that the policing of bathing was not always enforced. For example, writing under the pseudonym of 'lady bather's brother', an anonymous letter to the editor of the *South Coast Times* stated:

> I wish to again call attention to the fact that the privacy of the reserve for ladies at Wollongong is constantly pried upon by male creatures who evidently have no respect for themselves or ladies. Surely our worthy Mayor could get the police to move on this matter ... It is anything but pleasant for sensitive ladies to bathe before these peeping toms. (*South Coast Times*, 1908b)

This letter illustrated where the gender exclusiveness was broken by individuals not exercising self-control, the gaze upon the surf-bathing body was understood by 'lady bather's brother' as a sexual act. Positioned as 'peeping toms', the encounter between bathers and onlookers was defined in erotic terms. By breaking the gender exclusiveness regulations of the bathing reserve, men were robbed of their masculine valour and humanity and constituted as 'male creatures'. Solitary men who took pleasure in watching women bathe inverted the established middle-class social order. The peeping tom threatened the middle-class Victorian ideals of sexual restraint and marital, reproductive sex. Yet, in Wollongong, the letter to the editor suggested that masculine pleasure of watching women was not commonly regulated through punishment and policing of offenders.

Middle-class men visiting the beaches at Wollongong who drew on British bourgeois moral values also expressed in letters written to the editor their concern for women gazing upon the scantily clad surf-bather. For example, an anonymous letter to the editor of the *South Coast Times* in 1908, implored the authorities to enforce bathing ordinances to 'protect' female tourists from scantily dressed bodies at Fairy Creek bathing reserve in Wollongong:

> it is about time a little excitement was caused locally by causing local bathers to clothe themselves with decency, especially those who choose to bathe during the afternoon hours of Saturdays and Sundays. Twice in a week I walked past the gentlemen's bathing place with ladies who wished to visit Fairy Creek. On each occasion there were males disporting themselves in most meagre of loin clothes only, and made no attempt to make themselves less conspicuous than standing on the bank of the bath ... This part of the beach is one of the most attractive about your town and ladies should be able to visit it without being faced by men who evidently have no self-respect. (*South Coast Times*, 1908c)

The scantily clad surf-bathers visible to women were understood as socially and sexually threatening. This author was interested in protecting women from the sight of men dressed in the 'most meagre of loin clothes ... [who] made no attempt to make themselves less conspicuous than standing on the bank of the bath'. As White (2003) argued, the majority of those who argued against surf-bathing 'in public places' were those seeking to defend a Christian-infused morality that informed a British infused bourgeois masculinity.

Letters to the editor of the *Illawarra Mercury* also suggest that in the summers of 1908 and 1909, surf-bathers on Wollongong beaches had become a site of spectacle of masculine virility and cultural hysteria. The beaches of the Illawarra at this time became an increasingly charged site because of the rituals, behaviours and postures taken up by surf-bathers that did not conform to codes and conventions of respectable masculinity and lacked gender exclusiveness. As a letter to the editor in the *Illawarra Mercury* points out:

> But there are others who have an unsatiable (sic) desire to pose on the sands before an admiring crowd of both sexes, showing the symmetrical development endowed by a beneficent Creator, the pattern by which all men were designed. Three young men on the Bellinger River recently posed before some ladies in a similar way practiced in Wollongong for some years, and they got three weeks in prison. (*Illawarra Mercury*, 1908b)

According to this anonymous author, the surf-bathing body was gazed upon with admiration. The musculature was looked upon with a source of pride rather than anxiety. However, the concern was always that the regulatory frame of the pose and admiring gaze may slip into the erotic. Bathing ordinances were policies designed to encourage most individuals to exercise self-surveillance under the gaze of others. The anonymous author reminded readers that the bodies of men bathing in public were not to be presented to onlookers in ways to others that might be read as sexual or erotic.

Clearly, for some young men who regularly participated in surf-bathing at Wollongong beaches, an integral part of the pleasure of bathing was presenting their physically fit, toned and athletic bodies to be admired. However, it was exactly the masculine pleasure of being the object of a sexualised gaze that troubled the idea of bourgeois masculinity – and was an unmanly characteristic. As Maurizia Boscagli (1996: 170) has argued, 'the masculine pleasure in becoming the object of the gaze, removes this subject from the territory of the phallic masculinity into an abject space that bourgeois culture understands as feminine'. Bourgeois tourists may have had abject reactions to surf-bathers because of how they contested the gendered and classed corporeal borders of society. Tourists may have been both drawn to, and simultaneously repulsed by the spectacle of masculine virility displayed by surf-bodies at the beach. Newspaper articles certainly reported how some women were drawn to surf-bodies at the beach. For example, an article in the *Illawarra Mercury* reported that:

> Bathing was indulged in when hundreds of people were walking along the cliffs of a Sunday afternoon ... but now you found them – including

> a large number of ladies – on the cliffs watching the surf-bathers, and in a number of instances they had [looking] glasses. (*Illawarra Mercury*, 1909)

Clearly, in the summer of 1909, some women had a strong fascination towards gazing upon surf-bathing bodies. Yet, newspaper reports at this time were full of stories loathing the presence of surf-bathers. The surf-bathing bodies became a source of abjection because the border between the sexual and non-sexual boundary became permeable. Drawing on arguments of Iris Marion Young (1990), the surf-bathers may be understood as a socially abject group, or what she called 'ugly bodies'. Socially abject groups became a source of fear and anxiety because of how they unsettled the orthodoxy of the social order.

Tanning at the Beach: Sex-Crazed Perverts or Greek Gods and Goddesses

> Where a special sun-basking enclosure is provided the Council may erect notices prohibiting sun-basking except within such enclosure. (Clause 7 (a), Ordinance No. 52, Local Government Act 1906)

> Where such notice is exhibited persons shall not loiter or lie about the public bathing reserve clad only in bathing costume, except within such enclosure. (Clause 7 (b), Ordinance No. 52, Local Government Act 1906)

Tanning bodies were also increasingly disciplined through bathing ordinances. Throughout the 1800s, the imagined state of racial divisiveness of skin colour and respectable masculinity had worked against any desire amongst the colonial gentry to deliberately strip off and seek a tan by lying on the beach. In a highly racialised and classed society, 'savages' and 'labourers' had black and tanned skin, not those who aspired to the middle classes. Whiteness, throughout the 19th century, amongst the middle class in the British Empire, was taken according to John Urry (1990) as evidence of Englishness, delicacy, idleness and seclusion. One illustration of how tanning was embedded in a racialised narrative is a letter written to the editor of the *Sydney Morning Herald* by a woman using the pseudonym 'Daily Dipper'. She wrote that 'men taking sun-baths ... put themselves on the same level as dogs – well, blood will out' (*Sydney Morning Herald*, 1907d: 5). The equation of sunbathers with dogs was an overtly racist stance in the opposition to tanning.

In the first decade of the 20th century the practice of sun-basking or tanning was also associated with the emergence of surf clubs. Ancient Greece provided members of Sydney surf clubs with ready-made body philosophies of citizenship. In opposition to the indecency understood as rife at the beach, the founding members of Sydney surf clubs transferred some ideas from Ancient Greece and built a cult of the beach and of a natural, national, body that embraced tanning with zeal (see Chapter 5). These ideas were then circulated in the Sydney newspapers. For example, in 1907, the *Evening News* stated:

> Our Australian girls no longer consider it good to wear pale and uninteresting complexions ... these bronze Venuses, with ozone in their nostrils, and vitality in their constitutions, are to be the robust mothers of the vigorous race which is to hold white Australia against all comers. (*Evening News*, 1907, cited in Booth, 1991: 140)

This newspaper report reveals racialised and gendered truths as suntans purportedly transformed the bodies of women into 'bronze Venuses' and 'robust mothers'. Such articles worked within the conventional gendered assumption of women as mothers. However, becoming bronze, not white, was constituted as associated with gaining privileges associated with being more attractive, healthy and virile in comparison to an idle, delicate, feminine Englishness.

To help give legitimacy to the transformative possibilities of tanning, the surf lifesaving movement tapped into scientific discourses of heliotherapy and particularly the work of Dr Rollier of Leysin, Switzerland. Rollier's work was widely circulated in various medical institutions and demonstrated the scientific value of exposing the body to sunlight until a uniform tanning was produced in curing and preventing diseases (e.g. see Henry Dietrich, 1913; Roland Hammond, 1913). Furthermore, as Booth (2001), Daley (2003) and Grant Rodwell (1999) suggested, the regime of truth behind sunbathing and sunbathers was tied to eugenic ideals of tanned white skin. Bronzed bodies were an essential part of populating and defending the white Australian nation. The tanned body served to produce hierarchies of difference amongst Australians, with bronzed skin serving to signify different subject positions along these lines.

Yet, the zeal for the transformative possibilities of the act of tanning was not shared by the New South Wales Legislative Assembly. Instead, the bodies of sun-baskers were constituted as sexual, and therefore obscene. Like the dressing shed and bathing costume ordinances, the sun-basking legislation was also informed by thinking that framed sexuality in terms of Butler's heterosexual matrix and the gazing upon the genitals by a second party as sexual. The gender-exclusiveness of the sun-basking enclosures was another admission that the beach was a highly sexualised site. The New South Wales Legislative Assembly attempted to de-sexualise the beach by prohibiting the practice of sun-basking in bathing reserves, except where Council had provided two enclosures one for women, the other for men. Furthermore, to keep the bathing reserve as non-sexual, rules existed to restrict the time that surf-bathers clad in bathing costumes would be visible in public. Without a sun-basking enclosure, no surf-bather was permitted to remain dressed in their bathing costume for a longer time than was necessary to pass from the dressing shed to the sea, or sea-bath. Reminders of how surf-bodies were to move at the bathing reserve appeared regularly in newspapers. For example, at the start of summer in 1907, readers of the *Evening News* were reminded that surf-bathers were: 'required to take the most direct route between the dressing pavilions and the water' (*Evening News*, 1907, cited in Booth, 1991: 140).

In 1909, how shedding clothes and lying on the beach to sun-bask fell outside the sexual and moral norms of the bathing reserves in Wollongong was made evident after some Sydney municipal authorities provided sun-basking enclosures. Leone Huntsman (2001) discussed the pressure exerted by the Surf Life Saving Association in the construction of purpose built, single sex, sun-bathing enclosures at Bondi Beach. In Wollongong there was an outcry by aldermen. Alderman Krippner is reported to have said in March 1909 that sun-basking on Sydney beaches was 'not only demoralising, but disgusting' (*Illawarra Mercury*, 1909). For Alderman Krippner, sun-basking bodies at the beach were too sexy. He went on to say that he did not want bathing reserves of the Illawarra to 'fall into the same situation'. He regarded even the building of sun-basking enclosures on Sydney beaches as too risqué and inappropriate for respectable men and women. He was reported to have said that: '[i]f he stood alone he would try to prevent it [sun basking]'. Alderman Krippner concluded by remarking that:

> Before they [Wollongong Council Aldermen] granted the request [to sun-bask] they should be very careful, because it was far better to err on the strict side. (*Illawarra Mercury*, 1909)

Alderman Krippner's application of the cautionary principle to sun-basking drew on a regime of truth where civilisation was synonymous with a temperance that demanded the denial of the flesh. Similarly, the Christian infused morality behind the decision-making of the municipal authority informed Alderman Lance's thinking. He repeatedly argued that the indulgence in the sensual pleasures of both surf-bathing and sunbathing activities on 'the Lord's Day' was sinful (*Illawarra Mercury*, 1909), 'improper' and a 'desecration of the Lord's Day' (Bulli Surf Club records). To encourage individuals to exercise self-surveillance, readers were reminded that sun-basking was prohibited in bathing reserves; a decision that was to be 'strictly adhered to' (*Illawarra Mercury*, 1909). The majority of aldermen on Wollongong Council clearly believed that the bodies of sun-baskers were indecent. The decision to denounce sun-bathing taken by aldermen was informed by the belief that 'good citizens' remained fully clothed in public and did not flaunt their semi-naked bodies to the inspecting gaze of others.

Conclusion

By paying attention to the conflicting discourses that brought into existence surf-bathing bodies through bathing ordinances, new insights were provided to the importance of the spatial in conceptualising beach cultures. Surf-bathing bodies could not be separated from the ideas, meaning, practices and morals that shaped the space of the beach as a bathing reserve and in turn shaped the subjectivities of surf-bathers. This chapter has argued that an integral part of how spatial boundaries of the bathing reserve were continually made resilient and ruptured was around different sets of ideas or discourses that informed the practices and subjectivities of surf-bodies including fitness, sexiness, masculinity, femininity, movement, nakedness and clothing. Examining surf-bodies in the first decades of the 20th century in New South Wales was particularly interesting because the source of regulation over surf-bodies as too sexual – the Bathing Regulation Bill of 1894 – was increasingly disputed by members of the lifesaving movement in Sydney. In Sydney, during the first decades of the 20th century, the regulatory regimes of the lifesaving movement were crucial to mobilising an alternative gendered and racialised story of the bathing bodies at the beach to help imagine the exclusiveness of Australianness.

Unlike Sydney, however, the newspaper articles suggested that surf clubs in the Illawarra were not mobilised by discourses that framed the shape, skin colour, size, fitness and health of surf-bodies as the embodiment of Australian masculinity. There were no protests organised by Illawarra surf clubs at proposed skirted-bathing costume regulations or prohibited sun-basking. Amongst both municipal authorities and members of surf clubs the bodies of surf-bathers on Illawarra beaches were apparently not fashioned as essential to forging a new nation. Chapter 4 explores how aldermen in Wollongong became fixed on ideas of the coal and steel industries in imagining the Illawarra as Australia's industrial heartland. Chapter 5 discusses the lived experiences of members of Illawarra surf club members and their importance in reconfiguring and naturalising the patriarchal family.

Furthermore, the archive suggests that stories of the Illawarra beaches were dominated by the lack of self-surveillance and self-control amongst surf-bathers. Articles regularly reminded readers of the by-laws. Bodies of surf-bathers exhibiting no self-control were repeatedly constituted by aldermen drawing on biblical discourses as disgusting, indecent, and too sexual. For aldermen, bathing ordinances were therefore essential to sustain a social order over the beach. Similarly, those middle-class visitors who framed the semi-naked bodies on display in bathing reserves as too sexual sought reassurances of policing these by-laws because of the lack of self-control and restraint demonstrated by surf-bathing bodies. Despite the rhetoric of aldermen, the archive revealed little evidence of policing by nuisance inspectors of men who disobeyed the bathing ordinances. Newspaper articles convey the anarchy on the Illawarra beaches despite the regulations of the bathing reserves. It is possible to speculate why. On the one hand, some young visitors from Sydney perhaps took little notice of bathing ordinances drawing on ideas of Romanticism, primitivism and naturalism that fashioned the Illawarra beaches as a liminal space. On the other hand, Wollongong Council had little interest in enforcing bathing legislations because of limited financial resources, jurisdictions over several bathing reserves and a future envisaged in manufacturing industries.

3 Rail and Car Mobilities: Technologies of Movement and Touring the Sublime

In 1887 the railway line was completed between Sydney and Wollongong. Today, the 51-mile rail trip between Sydney and Wollongong seems perfectly ordinary, and is undertaken by hundreds of commuters on a daily basis. However, in 1887 the three-hour rail trip from Central Station, Sydney, for many signalled the securing for the Illawarra by the steam engine, a position as a seaside resort. Plans to extend the railway south to Kiama (opened in 1888) reinforced this position of the Illawarra as a holiday destination. For example, the *Sydney Mail* in an article titled 'Opening of the line to Wollongong and Kiama' reported:

> The Illawarra district is one of the prettiest and wealthiest in the colony and with communication from Sydney to the beautiful seaside resorts ... A great future should be in store for those places ... It is tolerably safe to assert that the day will yet come when the lovely beaches and fertile plains of the beauteous Illawarra district will be as familiar to Sydney people as are the many holiday resorts of Port Jackson. (*Sydney Mail*, 1888: 770)

With the railway connecting Sydney to the Illawarra, framed by discourses of the picturesque, the potential for Wollongong and Kiama to become holiday resorts for Sydneysiders was imagined as possible. This chapter investigates how first the steam engine and later the internal combustion engine of the automobile, boosted the Illawarra as a tourism destination in the social lives of men and women working in Sydney. Drawing on the work of John Urry (1990), this chapter explores how first the railway, and later the introduction of automobiles owned by a privileged few, transformed the spatial and temporal orderings of social life in the Illawarra. The railway and automobile technologies produced time–space compressions, where constraints related to space and time were progressively decoupled. With time and space no longer constraining movement, beach cultures of the Illawarra were opened up to be sampled and appropriated by more and more people living in Sydney. Drawing on familiar European Romantic Movement narratives dating from the early 1750s, brochures and guidebooks published by the New South Wales Government Railways and Tourist Associations fashioned the Illawarra as an escape from the frenzied pace of city living to an earthly paradise. The emergence of a tourism industry in the Illawarra is examined as one example of how State and corporate capitalism extended their material and cultural control by fashioning the boundaries of regions and nations. Through tapping into European Romantic moral geographies of sea air, water and scenic views the Illawarra was made visible as an earthly paradise, and made self-evident as a jewel of Australia:

> This is ILLAWARRA, a land fashioned for Fairies, to dwell upon; a spot smiled upon by Nature, in one of its generous moods.... Illawarra is beyond the power of painting. ('Unless to Mortals, it were given, To dip his brush, in dyes of Heaven'. *The Illawarra Tourist Guide: Wollongong Citizens Association*, c. 1920s, Foreword, writer and artist David Christie Murray)

The making of the Illawarra as a nature tourism and leisure destination cannot be understood apart from moral geographies; that is the values that guide actions and ideas about what is right or wrong, good or bad, better or worse (Smith, 2000). Tourist Associations of the Illawarra, since 1899, constituted and circulated a particular Australian version of the coastal 'idyll' in response to the social 'problems' arising from the rapid rise of industrialisation in Sydney. As David Bell (2006) outlined, the process of idyllisation is the outcome of how particular social groups appeal to specific sets of ideas that configure the places as distinct, separate and special. The coastal idyll is understood as the seaside of the city – constructed as a panacea by British bourgeois men for the supposed problems of the urban condition. Rapidly growing industrial cities were commonly understood by British bourgeois men as sites of physical, social and moral decay. Drawing upon Romantic ideologies, nature in the Illawarra became understood as offering restorative and regenerative possibilities despite the presence of coal mining. Hence, the Illawarra coastal idyll can be understood as Sydney's 'other' by drawing on the binaries of human/nature, urban/rural and self/other. For Sydneysiders, in response to the rapid growth of industries, the Illawarra became understood as a special place. In the Illawarra, moral geographies are illustrated through how nature was understood in terms of 'good' or 'healthy' sea and mountain air and how 'beautiful' views were understood to have restorative and regenerative possibilities for the body and soul.

Tourist guidebooks of the Illawarra of the early 20th century illustrate how the coastal idyll is a symptom of urbanisation. The coastal idyll is constituted as a panacea for the supposed problems of the industrial city. For example, in the pages of the guidebook, *The Beautiful Illawarra District, Wollongong. The Ideal Seaside and Tourists' Resort* (1910) the Illawarra is portrayed as idyllic through the earthly paradise trope:

> Heaven kissed the gentle vale when Nature spilled with lavish hand her benefits, her best treasures o'er this spot – the fairy tints and heavenly colours creating Earthly Paradise. And man came and called this land of Forest, Flowers and Ferns – this land of Glorious Natural Beauty – ILLAWARRA. (*The Beautiful Illawarra District, Wollongong. The Ideal Seaside and Tourists' Resort*, 1910, no pagination)

The coastal idyll is constituted through facilitating the promise of 'Glorious Natural Beauty', 'fairy tints' and 'heavenly colours'. With the Illawarra constituted as the 'land of Forest, Flowers and Ferns', the coal industry was silenced. Tempered by metropolitan middle-class sensibilities, the author went on to explain:

> Wollongong – the Queen City of Illawarra ... Set in one of Nature's fairest moulds, nurtured in environment pleasing to the most fastidious tastes, with the waters of the Pacific dancing at its very doors – sparkling, sun-tipped, ever restless, ever varying in their hue as the shadows fall aslant under the sun; glorious in its nestling dells and shady vales, with its background of undulating hills and noble mountain peaks, Wollongong to-day proudly rears its head as the Capital City of the district rightfully and lovingly designated – 'The Garden of New South Wales'. (*The Beautiful Illawarra District, Wollongong. The Ideal Seaside and Tourists' Resort*, 1910, no pagination)

The Beautiful Illawarra District, Wollongong. The Ideal Seaside and Tourists' Resort (1910) portrayed the dream of a coastal Arcadia. The emergent tourism industry helped to construct imagined geographies of Wollongong as the capital city of a garden. This understanding of Wollongong is not simply a product of metropolitan discourses, but retains traces of urbanity; it is imagined as a capital city. This kind of example suggests the inadequacy of thinking of the coastal idyll as purely an oppositional practice that appears to signal either 'human' or 'nature', 'urban' or 'coastal'. On the one hand, familiar discursive lines may sustain ideas and practices of coastal idylls. On the other, the discursive and material boundaries of the coastal idyll are never fixed, and always permeate each other within a dynamic social context. In effect, the process of idyllisation is comprised of familiar and unfamiliar threads of discursive and material practices woven together. Hence, by uncoupling the process of idyllisation from the normative prescriptions of the human/nature binary, it is possible to accommodate a multiplicity of different idylls.

The Steam Engine, Clock-Time, Industrialisation and the 'Tourist Class'

Tourism scholars and historians have concerned themselves with the social transformation of English bathing resorts for the elite into seaside resorts, catering for 'the people' (see Booth, 2001; Urry, 1990; Walton, 1983; Ward & Hardy, 1986). Together, these histories of the social transformation at English beaches at the turn-of-the-19th century tell a much larger story of how leisure practices were incorporated into men and women's aspirations for major social change – including improved working conditions, mobility and pleasurable lives. At the most basic level, this scholarship shows how the masses were mobilised to forge the 'tourist class' as a social collective.

Lash and Urry (1994), Mercer (1980), Urry (1990) and Waterhouse (1995) all agreed that changes in rhythms of the everyday life and places of the factory system were one of the most decisive moments for encouraging a collective social identity known as the 'tourist class'. Tracing the emergence of touristic practices, Richard Waterhouse noted how the spatial and temporal rhythms of industrialisation transformed the experience and organisation of social time and space:

> In pre-industrial Europe there was no clear demarcation between home and work, life and work ... Most labourers worked only for part of the year; bouts of intense work were followed, especially after harvest, by periods of idleness ... Holidays were determined by the Church calendar and were usually celebrated with collective activities – fairs, festivals and wakes. (Waterhouse, 1995: 4)

Waterhouse differentiated between the sense of time and place forged through the daily work rhythms and institutions of a European agricultural society in comparison to those experienced by the factory worker in the late 1800s. He demonstrated that the temporal rhythms of the factory system were central to the reorganisation of social time from the apparently 'natural' divisions of night and day, the seasons and movements of life towards death, to the mechanical. The regulatory rhythmic conventions of clock-time began shaping the lives of individuals and groups, including the precise timetabling of most work and leisure activities. The mix of social orderings of clock-time laid down spatial–temporal patterns that heightened the temporal and spatial differentiation between leisure and work.

Everyday understandings of clock-time as something to be saved, monitored, regulated and timetabled, facilitated the consolidation of tourism destinations by appealing to practices and experiences constituted as the antithesis of regulated and organised workplaces. Leisure time and spaces were understood as an 'escape' from the regulated, structured, monotonous and unchanging rhythms of workplaces. As Waterhouse suggested, tourism destinations were often pitched by the emerging tourism industry as an 'escape' back to 'nature' or promising 'freedom' from the responsibilities and regulation of clock-time.

In the mix of social ordering of clock-time were State processes that legislated for official breaks. In Britain, the 1833 Factory Act enforced two full days and eight half-days of holidays per year. Later, in 1871 in Britain, the legislation for the Bank Holiday Act was passed – a culmination of the holiday with pay movement. The Act stipulated fixed public holidays at Christmas, Easter, Whitsunday and the first weekend in August.

In the Australian colonies, during the 1850s, the Eight-Hour Labour Movements marched for shorter working days and official holidays. This movement drew support from Dr Thomas Embling, a medical doctor from the United Kingdom, who is credited with coining the slogan '8 hours labour, 8 hours recreation, 8 hours rest'. He argued that increased hours of leisure would give workers the opportunity to become 'healthy, wealthy and wise' (Clark, 1978: 93–94). In 1856, the direct action of Stonemasons in Melbourne successfully achieved the introduction of an eight-hour day within the building industry, without loss of pay (Lynch & Veal, 1996). However, in New South Wales at the end of the 19th century the whole labour movement was at a low ebb because of economic depression and drought (Hogan, 2008). In the Illawarra, by the 1870s, the first trade unions had consolidated around the working conditions of coal mining companies. It was not until after Federation and further campaigns by trade unions that an eight-hour working day was achieved nationally in the 1920s.

The Illawarra or South Coast Railway

In the Illawarra, as elsewhere in the late 1800s, the railway was exceptionally important in fashioning tourism geographies by transforming the relationships between time and space. As Urry (1990) has pointed out, the railway created its own geographies through changing the material, social, economic and embodied relationships that comprise place. At one level the railway created pulses and flows at stations through the spatio-temporal patterns of the timetable. These pulses and flows became coordinated throughout the world by around 1884 following the adoption of a standardised international clock-time, Greenwich Mean Time (GMT). At another level, the extraordinary mechanical power of the steam engine seemingly drew those places connected to the rail network closer together in time and space, and further distanced those not linked by rail. As David Harvey (1989) argued the railways appeared literally to compress time and space. The 51-mile journey from Sydney to Wollongong took two to three days by horseback at the beginning of the 19th century and just over two hours by rail at the start of the 20th century. While the railway reconfigured travelling to the Illawarra as a day-trip from Sydney, the failure of the railway to ever reach Bega and Eden fuelled discourses of the 'forgotten corner' (of New South Wales) and the 'Far South Coast'. The introduction of new modes of transport enabled people to compare and contrast how places were networked and on occasion to express as a regional identity the virtues and problems of slower ways of overcoming distance. To better understand how the train democratised travel and fuelled mass tourism to the Illawarra we need to explore the mechanisms that brought this about.

The excursion ticket and 'special trains': Popularisation and democratisation of rail travel

In Britain, Thomas Cook, a mission worker for the Baptist Church, alongside railway companies, started organising railway excursions from the early 1840s that democratised longer-range travel. In particular, the day and week excursion railway ticket brought prosperity and mass tourism to the beaches of England. For example, in 1841, the London–Brighton railway opened. Brighton visitor numbers subsequently increased from 7000 in 1818, to 65,000 half a century later (Ward & Hardy, 1986). With the 'tourist class' becoming centre stage of certain commercial enterprises during the second half of the 19th century, English seaside resort tourism was characterised by the building of infrastructure such as entertainment piers and hotels. For example, in Brighton, the West Pier was built between 1863 and 1866, followed by the Palace Pier (1891–1899); the Grand Hotel opened in 1862, followed later by the Metropole Hotel (1888).

Likewise, excursion tickets were integral to the growth of mass tourism from Sydney to the Illawarra but not until the 1920s. The New South Wales

Government Railways (1855–1972) played a central role in promoting the Illawarra as a mass tourism destination from Sydney through sales of excursion tickets during the first decades of the 20th century. As was the case in England, excursion fares were particularly useful in increasing flows of tourists through the democratising of travel. William Bayley (1975: 57) positions the Illawarra during the 1920s as 'Sydney's great southern coastal playground.'

Excursion fares popularised and cheapened travel, and were promoted as a 'single fare for the return journey'. They included an open return fare, the return journey occurring within one month from the date of issue (New South Wales Government Railways passenger fares and coaching rates, 1920, 1922). Excursion fares were promoted in the local print media as well as in tourism brochures and guidebooks published by New South Wales Government Railways. For example, an *Illawarra Mercury* advertisement stated:

> Excursion Tickets at a Single Fare for Double Journey (minimum rates 8 shillings/9 pence First Class and 5 shillings/10 pence Second Class) will be issued at Wollongong to Sydney and intermediate Stations by train leaving Wollongong at 8:48 a.m. on FRIDAY, 1st November ... Tickets will be available for return within one calendar month. (*Illawarra Mercury*, 1929a)

Similarly, the New South Wales Government Railways' brochure, *By Train in Daylight through the Beautiful Illawarra* (c. 1930, no pagination) promoted the excursion fare explaining: 'a return ticket costs only the usual single fare'. Affordability and flexibility were essential elements in the Railway Department's promotion and generation of mass tourism. Travellers were not only given cheaper fares but greater flexibility.

'Relief', 'special' and 'excursion' trains were integral to mobilising tourist flows. From the 1920s the New South Wales Government Railways' archival records included notices of 'special trains', 'special relief passenger trains' and 'excursion trains'. These notices were regularly sent out to all stations as well as to guards and train drivers. 'Special' excursion trains were additional trips to the 'normal' timetable and were organised by New South Wales Government Railways during the summer holidays to promote train travel to specific destinations, including the Illawarra. For example, in the Christmas holiday period of 1920, notice is given of additional 'special excursion trains to and from the country' (Special Train Notices, 1920). Throughout the 1930s the railway archives reveal further evidence of the so-called 'relief' and 'excursion trains'. For instance, 'Special Train Notices' for the 11th and 18th of January, 1930 announced that there were to be 'relief passenger trains' from Sydney to Kiama at 9:10 am on Saturday morning and 'tourist trains' on the Southern Line (travelling to the Illawarra) were to commence in February (Special Train Notices, 1930).

Special or promotional events of the New South Wales Government Railways were of equal importance in the genesis of mass vacationing to the Illawarra. For example, on the 20th October 1930 there was an employees' picnic held at Nowra, and a special Sydney to Nowra 'picnic train' scheduled to cater for the occasion (Weekly Special Train Notices, 1930). In 1936, there is a record of a New South Wales Government Railways sponsored train excursion to Wollongong. An article in the *Illawarra Mercury* titled, 'Coast Trip, Tablelands to Wollongong' states:

> One of the most enjoyable tourist outings in the history of the tablelands was enjoyed by over 750 residents who took advantage of the train arrangements on Sunday last. (*Illawarra Mercury*, 1936a)

According to the report in the *Illawarra Mercury*, two trains, one from Bowral, the other from Mittagong, showed visitors 'delightful views', while the 'North and South beaches ... were favourites throughout the day' (*Illawarra Mercury*, 1936a). In 1937 a 'cruising train' visited the Illawarra and the Shoalhaven 'conveying a party of 132 [school] girls' (*Illawarra Mercury*, 1937a). Readers are informed that 'another party of school girls will arrive to-day and take the same trip'. The following month there were further reports of 'railway excursions' on offer, which would coincide with 'the opening' of 'the surfing season' and include 'many popular tourist resorts ... along the coast' (*Illawarra Mercury*, 1937b). These records illustrate how the novelty of rail mobility, as much as the destination, was considered as a tourist resource. The journey was made exciting and worthwhile by travelling through a landscape understood in British terms as offering 'delightful views'.

In the 1930s, these sponsored trips did not go without praise from various surf clubs. In 1937, the Wollongong Surf Club commended the Railway Department, reporting that:

> We had a good crowd in the water and the reserve in front of the club house was thronged with picnic parties. It was due to the enterprise on the part of the Railway Department, which sponsored such a successful outing for the metropolitan residents. (*Illawarra Mercury*, 1937c)

Austinmer Surf Club wrote two months later, again extolling the Rail Department:

> The Railway Department provided special excursion trains to Austinmer on Sunday, and judging by the thousands that came down and patronised the surf and beach, the excursions proved a huge success. (*Illawarra Mercury*, 1937d)

On this particular occasion the Railway Band from Sydney played music to enhance the beach as a site of spectacle during the surf lifesaver march past and musical flags.

Excursion fares and the so-called 'special trains' of mass tourism transformed the Illawarra beaches and played a crucial role in securing funds for the railway. The 1940 Annual Report on Traffic Branch Operations stated that there was a 'permanent demand' for sponsored promotional travel to the Illawarra. Between 1939 and 1940 the New South Wales Government Railways' records suggest that sponsored trips alone accounted for '22 trains, 9,841 passengers, and £3,604 in revenue' (Annual Report on Traffic Branch Operations, 1940). 'Relief', 'special', 'cruising' and 'excursion' trains also enabled a particular type of mass tourism in the Illawarra. Unlike Brighton, England or Long Beach, California, there were neither dance-halls nor amusement parks at the seaside resorts of the Illawarra. Instead, the railway was central to orientating commentary and defining the Illawarra as both a leisure and nature space. Excursion trains facilitated access to a number of leisure activities including sightseeing, camp sites and promotional events often organised by the New South Wales Government Railways including picnics, 'mystery' hikes, and later, surf carnivals.

Sight-Seeing: A Rail Trip through an Earthly Paradise

> The Illawarra or South Coast Railway is essentially a tourist line. No other passes through so much beautiful and romantic scenery within so short a distance of the metropolis, and it is questionable whether any other section of the Railway system comprises so many miles of unbroken charm and continuous interest. The characteristics of the landscape belong to this particular railway journey, and are not within the scope of any other line, lying full within its track also being that narrow strip of country that admiring generations have agreed to call The Garden of

> New South Wales. (New South Wales Railway Holiday Resorts, cited in *Beautiful Illawarra: The Illawarra or South Coast Tourist Guide*, 1899: 5)

Since the opening of the Illawarra or South Coast Railway, the New South Wales Government Railways Commissioners grasped the promotional opportunities of the coastal journey as a tourist attraction. In the first decades of the 20th century, the sight-seeing pleasures of train mobility remained mediated by European Romantic representations in art and in advertising of the coastal and mountain idyll. For example, over a decade after the opening of the South Coast Railway, an article in the *Sydney Mail* drew on narratives of the 'picturesque' to convey the pleasures of travelling by train through the Illawarra:

> the coastal trip to the fruitful south is highly picturesque, and while abounding in pretty glimpses of tangled foliage, lordly trees, and beautiful ferns, on the one side it rises to the magnificent, when . . . at the turn of a curve brings the broad, Pacific into view. (*Sydney Mail*, 1906b: 620)

Guidebooks and brochures marketing the South Coast Railway journey also provided detailed instructions on how to anticipate and appreciate the sparsely settled trees, ferns, coasts and mountains whirling by the window. Rail travel to the Illawarra was framed by a particularly Australian version of a coastal and mountain idyll. Pitched as a journey through the 'The Garden of New South Wales', tourist brochures commissioned by the New South Government Railways reproduced gendered, classed and ethnic assumptions about the Illawarra first brought by British and European colonial travellers, painters and explorers in the 1800s (see Chapter 1).

For example, the brochure titled, *By Train in Daylight through the Beautiful Illawarra* (c. 1930) published by the Commissioner for Railways, New South Wales, is an illustrated description of what is pitched as an 'enchanting journey', referring to many 'tantalising spots' and tells of the 'kaleidoscope sequence of unsurpassed views'. Urry (2000: 56) termed this way of seeing landscape as 'panoramic perception'. The South Coast Railway journey was considered special because it enabled travellers to witness with their own eyes the rainforest and the beach positioned as scenic 'virgin forest and ocean expanse' (*By Train in Daylight through the Beautiful Illawarra, c.* 1930, no pagination) (see Figure 3.1). Tourists were encouraged to view 'scalloped seashores and placid inlets, towering mountains and sloping meadows'. Of Stanwell Park the brochure stated:

> For some distance the railway skirts the rim of a huge amphitheatre . . . hundreds of feet below, the long Pacific rollers are seen chasing each

Figure 3.1 South Coast Railway, Stanwell Park (1906) (*Source*: unknown. From the collections of the Wollongong City Library and the Illawarra Historical Society). Stanwell Park Beach is visible in the top left-hand corner of the photograph

> other in ceaseless procession through a delicate tracery of foam and expiring on the golden sands. (*By Train in Daylight through the Beautiful Illawarra*, c. 1930)

The brochure gendered the landscape as feminine, illustrated by references to 'virgin forest', 'placid inlets' and the 'delicate tracery of foam'. The tourist is assumed to be male. The Illawarra landscape, gendered as feminine, reproduced taken-for-granted assumptions about the capacity for men to visit and explore. It was seemingly men who were assumed by the railway marketing authorities to be making the decisions about travel in the 1930s. New South Wales Government Railways' marketing discourse was integral to circulating and reproducing particular gendered ideas of the Illawarra landscape as a travel destination. As signalled in this brochure, the South Coast Railway generated a new, pleasurable way of appreciating the coast, and was an important way in which British Australians could experience myths of sparsely populated coast lines beyond the metropolitan centre as a tonic to the ills of the city. As Glenn Mitchell (1997: 145) argued, for many British Australian Sydneysiders in the 1930s, rail mobility allowed the

Illawarra to be imagined and experienced as the 'Jewel in the Crown of Leisure'. However, it is important to remain mindful that obscured by commercial imperatives and promotional stories of mass consumption, the land belonged to someone else.

Picnicking

> Stuart Park, Wollongong. Nature was in one of her most pleasant moods in forming this beautiful spot ... there is a carriage drive on its banks one mile in length which commands an ever-changing view of sea, lake and mountain views, in some parts traversing, avenue fashion, belts of primeval bush, forming on its banks grassy glades for tourists or picnic parties. (*Beautiful Illawarra*: *The Illawarra or South Coast Tourist Guide*, 1899: 57)

From mid-19th century and throughout the early 20th century, one of the dominant visitors to Illawarra beaches was picnickers. As discussed by Cameron White (2006), picnicking was a very popular middle-class colonial leisure activity in Sydney. Popular picnic sites for those taking day-trips in Sydney included Watson's Bay, South Head and Bondi Beach. The earliest Illawarra guidebooks recommended a number of beaches for picnicking including Thirroul; Austinmer; Stuart Park, Wollongong; Long Point, Shellharbour; Gooseberry Island, Lake Illawarra; Seven Mile Beach Gerringong; and Ourie Beach, Gerringong. The popularity of the Illawarra beaches for picnickers can be in part explained by how the practice was informed by the demand for sites that would provide views understood within the Romantic tradition as beautiful or picturesque. For example, *The Illawarra or South Coast Tourist Guide* (1903) explained:

> Stuart Park, Wollongong ... represents one of the beauty spots of Wollongong, and perhaps the ideal Park of the State ... Romantically situated on the Shores of the Ocean in a sheltered spot close to a splendid Estuary, upon which boats are placed ... Here the toils and troubles incidental to life are banished away amid the splendour and grandeur of the surroundings ... The Pacific Ocean on one side – an object lesson from Creation – attractive scenery immediately adjoining ... no wonder tourist and dwellers find the Park a glorious place to linger in. (*The Illawarra or South Coast Tourist Guide*, 1903: 38)

In this case Stuart Park, Wollongong was praised as 'Romantically situated' and offering 'attractive scenery'. Guidebooks located shaded, secluded and picturesque vantage points in 'nature' where picnic blankets could be laid out and the beauty of the panoramic view appreciated over food. Thus, as argued by White (2006), the fundamental logic of the picnic relies upon reproducing the fundamental colonial myths of *terra nullius* (nobody's land). The beach reinvented by middle-class British colonists as picturesque backdrops to picnics, was taken out of the realm of home for Aboriginal Australians. The myths that sustained the picnic helped displace recognition of the displacements and dispossessions of Aboriginal Australians.

Stuart Park, framed as a site offering access to 'primeval' nature fulfilled the fundamental Romantic logic of an escape from the city, to a place understood as aesthetic and untouched by humans. Possibilities were offered through the picnic to express social distinction, rejuvenation and self-discovery through picnicking. For picnickers, framed by the European ideology of Romanticism, the importance of the beach as a picnic spot was as a panoramic view for a fantasised encounter with beauty, rather than bathing. Indeed, as stressed by White (2003), a crucial aspect of picnic etiquette was that no-one bathed. Instead, particularly valued picnic spots in tourist guides, like Stuart Park, often provided facilities for boating and fishing as well as swings and sports ovals. In the early 1900s, the bodies of picnickers affirmed hegemonic constructions of gender for middle-class men and women by not bathing (Figures 3.2 and 3.3).

Hiking

Men from bourgeois society had long assumed the right of access to the Australian bush as wilderness, through the establishment of bushwalking clubs. Wilderness had long provided a screen on which to project a particular classed version of conservation and white masculinity through 'campcraft' and 'bushcraft' skills. However, to attract women and men, the leisurely walking practices facilitated by the South Coast Railway tapped into different sets of ideas derived from bodily rejuvenation and moral transformations through clean air, sunlight and views from mountaintops.

The New South Wales Government Railways benefited from the New South Wales Federation of bushwalking clubs that emphasised the moral benefits of leisurely walking along paths that had already been walked, through wilderness; imagining bushwalking as a cultivating experience that rejuvenated the individual, strengthened the body and soul and forged an Australian identity through an appreciation of landscapes, seascapes and

Figure 3.2 'On Thirroul Beach' (c. 1910) (*Source*: Alegernon Winn Collection. From the collections of the Wollongong City Library and the Illawarra Historical Society)

Figure 3.3 'A Beach Scene Thirroul' (c. 1910) (*Source*: Alegernon Winn Collection. From the collections of the Wollongong City Library and the Illawarra Historical Society)

panoramas that entailed slowness and effort (Harper, 2007). Like picnicking, bushwalking was another way British colonisers could claim authority over Aboriginal country. For example, from the late 1920s, Illawarra guidebooks were creating anticipated landscapes for tourists, and providing instructions on where to walk and how to gaze:

> Austinmer: The Mountain climb to the world-renowned **Sublime Point** (1,330 feet high) up the ladder of steps rewards the tourist with a Panoramic view of landscape and seascape unsurpassed in any part of the world. (*Wilson's Rail, Road and Sea Guide to the South Coast and Southern Highlands, New South Wales*, 1929: 5, bold in original)

Drawing on discourses of the sublime, the mountain climb in Wilson's guidebook was positioned as offering 'unsurpassed' views. Appreciating the seascapes and landscape requires physical effort, climbing 1300 feet up a series of steps. Similar ideas are circulated in an article in *The Sun* titled: 'Hiking: The thing that spells freedom in the true sense of the word', that reported a 'mystery hike' organised by the New South Wales Government Railways to Stanwell Park:

> Freedom in hiking is found among laughter-loving young people ... makes one feel free, to run and shout with the pure joy of being able to lift your voice to the clear blue heavens above and hear its ringing echoes from the opposite side of some rocky gully ... Hiking fosters the worship of sunshine, fresh air, outdoor beauty, physical fitness ... healthy young bodies and minds. (*The Sun*, 1931: 23)

The view of mass hiking expressed in this newspaper article served to redraw a sharp line between the 'city' and the 'bush'. Set against a wilderness backdrop, hiking was cast as freedom for young people from the city and work. Hiking bodies were legitimised within medical–moral discourses as a panacea for the supposed problems of the urban condition. The Illawarra becomes storied through hiking through wilderness; praised as a fine Australian tradition of nature conservation and improvement of the self through combining physical exercise with the experience of wilderness, while writing out Aboriginal Australian connections to country.

The railways enabled the popularisation and democratisation of walking during the 1930s, with events including over 8000 hikers. According to an article in *The Sun*, by August 1932, approximately '53,000 hikers had contributed £160,000 towards railway revenue' (*The Sun*, 1932: 8). Accessible by train, and framed by tourism guides as offering experiences of both the mountain and coastal idyll, the bush paths constructed through the Illawarra Escarpment, including the mountain climb to Sublime Point, became popular destinations for mystery hikes organised by the New South Wales Government Railways and the independent tourist.

Camping

> Stuart Park ... Situated close to the seashore in North Wollongong, Stuart Park is an extensive camping ground, providing special facilities for those who spend their holidays under canvas. The camping space is divided into allotments which are let by the caretaker living in the Park. The caretaker also conducts a kiosk, supplying refreshments and general provisions, while local tradesmen call daily at the Park during holiday periods to deliver bread, meat, milk, fish and other requirements. (*Picturesque Illawarra: The Garden of New South Wales*, 1931: 5)

In the Illawarra during the first decades of the 20th century, recreational camping was an expression of both the shortage of accommodation, and automobility and mass tourism enabled and encouraged by the New South Wales Government Railways. From the 1920s, municipal authorities established camp sites adjacent to bathing reserves patrolled by surf lifesavers including Coledale, Bulli, Corrimal, Fairy Meadow, Stuart Park, Shellharbour and Kiama (Figure 3.4).

Recreational camping sites in the Illawarra were erected drawing on ideas of predecessors in Britain, but were often selected at former Aboriginal camping sites. During the 1920s a range of recreational camping movements flourished in England, including the Holiday Fellowship and the Youth Hostels Association. As a low-cost practice, recreational camping was legitimised by medical discourses circulating about the individual health benefits of sunshine, sea air and salt water. In England, Ward and Hardy (1986) outlined how a generation had grown up with the interweaving of health benefits and camping through the moral geographies of seaside camps run as early as the 1870s by philanthropic organisations for the working-class poor. Camping followed the pattern of much open-air activity organised by philanthropic organisations at this time (Rosenthal, 1986). Camping-in-nature was a valued way of dwelling for social reformers to maximise exposure to the regenerating and purifying qualities of air, water and sunlight. Nature at the seaside was believed by philanthropic

Figure 3.4 Camping at Stuart Park (c. 1925) (*Source*: unknown. From the collections of the Wollongong City Library and the Illawarra Historical Society)

organisations to have socially transformative qualities for the urban working class.

The moral geographies of seawater and sea air configured as a tonic for the ills of the city were not lost on the New South Wales Government Railways. This version of the beach was promoted in brochures. Camping advertisements encouraged people to camp through a combination of excursion fares and the configuration of the beach as scenic landscape. A New South Wales Government Railways brochure, circa 1930, advertised, beside a photograph of a beach:

> one of the prettiest spots one could wish to see. You may camp here and spend the holiday you have always dreamed of. (*By Train in Daylight through the Beautiful Illawarra*, c. 1930, no pagination)

This brochure tapped into the demand for the pleasures of going away to the beach, mimicking desires for the seaside in Britain.

In the Illawarra, during the 1930s it is hard to over-emphasise the popularity of beachside camping. The surf club records provide an indication of how surf lifesavers interpreted the inflow of summer campers. For example, in 1937 writing under the pseudonym, 'Surfo', a surf lifesaver from Wollongong Surf Club noted:

> The holidays are gone again, and with them most of the large crowd of campers who were assembled in Stuart Park, and had many an enjoyable dip in the surf at our ever-popular beach. (*Illawarra Mercury*, 1937e)

And a Bulli Surf Club report in the *Illawarra Mercury* noted that for the summer of 1939:

> Campers are realising the excellent conditions under which they can camp and certainly this year must have eclipsed all records as regards campers. (*Illawarra Mercury*, 1939a)

Similarly, an article in the *Illawarra Mercury* for the previous summer of 1938 commented that:

> The campers were in evidence in proximity to the beaches, and in the majority of cases were loud in their praises of the sites selected. It appeared to be the consensus of opinion amongst the visitors, that the Illawarra district holds attractions for tourists that are not to be found elsewhere – wonderful surfing, delightful scenic resorts, combined with all the amenities of present-day life. (*Illawarra Mercury*, 1938a)

According to Bayley (1975), in the 1930s, the campsite at Bulli Beach could accommodate at least 1000 tents. However, unlike Britain, the demand in the 1930s did not trigger investment into the holiday camp that facilitated seaside holidays particularly for nuclear working families pitched as a tonic for the city. Orvar Löfgren (1999) and Colin Ward and Dennis Hardy (1986) have written about the importance of the opening of Butlin's Holiday Camps at British coastal locations in the 1930s in the process of reinventing and repositioning the beach as the normative location for a 'family holiday'. This work also revealed how class differentiations become rendered visible through certain middle-class narratives of holiday camps. In England, holiday camps for lower socio-economic groups became portrayed as the objects of class censure by calling upon working-class stereotypes as sexually excessive, abject and immoral. In the Illawarra, it was the history of the coal and steel industry, rather than that of the holiday, through which the dominant definitions of the moral geographies and class identification with the region occurred (see Chapter 4).

Automobiles and Reconfiguring the Coastline as an Earthly Paradise

> The **Bulli Pass** and **Sublime Point**, about one hour's drive along a pretty road, if not visited before, is, of course, one of the chief attractions. We don't attempt to describe its charms; they are too well known. (*Wilson's Rail, Road and Sea Guide to the South Coast and Southern Highlands, New South Wales*, 1929: 77, bold in original)

As automobiles became increasingly affordable and reliable through mass production in the early 1900s, automobile mobility generated new industries and objects to sustain the pleasures of motor touring – including car hire (with driver), road signs, signposts, guest houses, a national road building program, guidebooks and tour guides. Georgine Clarsen (2008: 121) outlined how the arduous automobile journey 'had long been central to Australia's status as both a colonized and colonising European outpost'. She outlined how after World War I in Australia transcontinental motorists crossing trackless deserts emerged as national heroes, 'conquering' what was assumed to be a male frontier in the popular press. Rather than the rigours of travel in central and northern Australia that lent to the formation of heroic stories, Tourist Associations in the Illawarra emphasised the luxury and comfort of travel through nature as scenic to help facilitate the formation of upper- and middle-class identities.

Tourist Associations in the Illawarra seized upon promoting the commercial opportunities that automobile journeys provided as day-trips from Sydney. For example, how motor touring spawned innovations in the tourism industry is evident in an advertisement that appeared on the front page of the *South Coast Times*:

> Motor cars for hire, trips arranged to Illawarra Beauty Spots, whole day or half day trips. First-class Driver. Luxurious Car. Comfort assured to patrons. Book early. Picnic Hampers supplied with all leading brands of Liquors, Spirits and Wines. (*South Coast Times*, 1914b)

Furthermore, as evident in this advertisement, the Illawarra day-trip configured motor touring not to avail itself of heroic stories. Instead, drawing upon ideas from Britain, these motor tours were a form of class distinction, based on the concept of the display of social status through an appreciation of the 'open road' at a slow meandering pace, with stops along the way to appreciate 'beauty spots', food and drink. Hence, in this advertisement emphasis was given to finding pleasures in the luxury and comfort of the travelling conditions and the quality of the picnic refreshments. In the 1920s, the Illawarra became a scenic backdrop to motor tours through which Sydneysiders could demonstrate aesthetic taste and class distinction.

The car's significance in changing touring practices grew as social life became reconfigured through interconnecting highways. For instance, the completion of the Prince's Highway in the 1920s as part of a national road building program was important in how the Illawarra become repositioned in what Urry (2000: 59) termed an 'automobilised time-space'. In motor touring guidebooks of this time, the Illawarra became a 'beauty spot', one of many 'must see' scenic attractions along the coastal route of the Prince's Highway from Sydney to Melbourne. For example, in 1929 an article in the *Illawarra Mercury* praised the publication of the *'Official Prince's Highway Motor Guide'*:

> Equipped with the Guide, the user must feel that sense of security and appreciation which attends a personally conducted tour as the various beauty spots and places of interest are located by the speedometer while driving. Hotels, Guest Houses and Garages are listed at each town en route, while where facilities are available, camping grounds are specified. (*Illawarra Mercury*, 1929b)

Similarly in 1931, 3000 copies of a motor guide (*Picturesque Illawarra: The Garden of New South Wales*) was reported in the *Illawarra Mercury* as a 'notable achievement of the year' (*Illawarra Mercury*, 1931a). This motoring guidebook contained a road map of the Illawarra, an accommodation directory and a list of scenic attractions and was a joint publication of the Illawarra Tourist and Publicity Bureau and Illawarra Motor and Tourist Services.

Drawing on Urry (2000) these guidebooks draw attention to 'two interdependent features of automobility: that the car is immensely flexible *and* wholly coercive' (Urry, 2000: 59, italics in original). The car became a source of freedom for motor tourists through the flexibility it provided in terms of controls over time, speed, distance, where to stop, and what to visit. Cars extended when and where people could visit. However, at the same time, touring by car became wholly coercive. Guidebooks like the *Official Prince's Highway Motor Guide* were produced to orchestrate the flow of tourists. Guidebooks overcame the uncertainties that freedom of the road brought by coercing people to stay, stop, look and appreciate certain sites along the route.

Motor touring through the Illawarra was constituted not only as a privileged and pleasurable activity but also as how to appreciate the landscape and the nation. Consequently, motor touring guidebooks of the Illawarra not only helped to popularise certain roads as 'tourist routes' but also gave instructions where to stop and how to appreciate the ocean and beach.

For example, views from Sublime Point and The Bulli Pass Lookout (Figure 3.5), on the Illawarra Escarpment were pitched in guidebooks as an essential stop en route between Sydney and Melbourne on the Prince's Highway. The guidebook, *Picturesque Illawarra: The Garden of New South Wales* (1931) explained that:

> The Prince's Highway and other roads enter the district by several mountain passes which wind through the beautiful Australian native bush, and so it is that the motorist, before descending one of these passes, is confronted by a panorama of outstanding beauty, a blended picture of seaside, bush and mountain scenery. In particular, the views from Sublime Point and Bulli Pass Lookouts have been lavished with expressions of admiration by Australian and Over-seas visitors. (*Picturesque Illawarra: The Garden of New South Wales*, 1931: 3)

The guidebook recommended stopping at Sublime Point and the Bulli Pass Lookouts. Here motor tourists could admire the scene laid out before them framed as a 'panorama of outstanding beauty'. Views from the top of the escarpment were particularly valued because they combined the coastal,

Figure 3.5 'View from Bulli Lookout, Bulli Pass, Illawarra Range' (c. 1920) (*Source*: unknown. From the collections of the Wollongong City Library and the Illawarra Historical Society)

mountain and rural idyll. In 1928, a new kiosk was built at the Bulli Pass Lookout. Men and women could sip tea while enjoying the panorama.

Picturesque Illawarra: The Garden of New South Wales (1931) not only explained where to gaze, but how to gaze. For example, the following poem by Michael Hennessy was published in the inside cover of the motor guide:

I stand upon a lofty cliff
Beneath me I behold
The Illawarra beautiful
With surf and sands of gold

Lagoons and lakes and many towns
Spread out before my eye,
And on the right, a mountain range
Rears high into the sky.

Australia has been often called
A jewel set in the sea –
I know exactly at what point,
It shines most brilliantly (M.P.H.)

Picturesque Illawarra: The Garden of New South Wales (1931) suggests how motor touring framed the Illawarra as an earthly paradise by providing access to certain vantage points. Steeped in aesthetics of Romanticism, this poem validates the mastery of the eye over the country and confirms a theme at the heart of national life – the sea brings symbolic unity to the borders of Australia. The pleasures of motor tourists visiting the Bulli Lookout play a part in shaping understandings of the Australian nation, conferring a sense of belonging even through a brief visit.

Furthermore, in the 1930s, motor guides were still directing tourists down the Bulli Pass to view the 'beautiful Australian native bush' (Figure 3.6). Visitors were guided to see exquisite scenes because of its abundance of tree ferns, tall trees, vines, creepers, palms and wildflowers. Indeed, motor tourists, dressed in their best clothes, stopped and picnicked along the roadside of the Bulli Pass to appreciate the aesthetics of the rainforest. As noted by Julia Horne (2005), the fern was well established as a symbol of middle-class taste in Australia.

By the mid-1930s motor touring had helped secure the Illawarra, competing favourably with the Blue Mountains, as a tourism destination from Sydney. The Blue Mountains offered visitors from Sydney primarily leisure activities built upon an Australian version of the European wilderness myth that sustained cultures of nature as scenic and a remedy to the alleged social ills of urbanisation and industrialisation. At a Wollongong Tourist Association

Figure 3.6 Bulli Pass (c. 1935) (*Source*: unknown. From the collections of the Wollongong City Library and the Illawarra Historical Society)

meeting in September 1935, Mr Cahill spoke of how motor touring to the Illawarra fulfilled fantasies of journeys made by city drivers for both mountains and the ocean. Of the 'scenery along the Prince's Highway', he is quoted in the *Illawarra Mercury* as saying:

> it was without a rival in the State, even the Blue Mountains could not approach it. It has all the Mountain attractions, with the addition of the blue Pacific in the foreground. Some of the scenic beauties reminded one of the scenic gems of England. (*Illawarra Mercury*, 1935a)

As evident in this quotation, motor touring also worked in particular ways in Australia to inscribe British understandings onto the bush. The shared culture of motor touring helped naturalise the presence of British settlers in an Australian wilderness. Stories told about motor touring by British settlers were therefore often about the role of automobility in establishing a sense of self and belonging in a new nation.

Conclusion

The railway and the automobile fashioned the seaside resort in the Illawarra in similar and different ways to Britain. Similar to Britain, the railway was integral to the personification of a new tourist class, mobilised by low-cost fares and the organisation of mass excursions. The railway expanded the social capability of who could afford to travel from Sydney to the beaches of the Illawarra through generating a time–space prism fashioned by timetables, steam engines and stations. Like in Britain, the railway was integral to triggering mass tourism and the identity of the tourist. In Australia, the railway was also crucial to inscribing colonial meanings onto places. The New South Wales Government Railways was integral to revitalising familiar British narratives of nature, to frame the Illawarra as an earthly paradise – often described as 'scenic', 'beautiful' and 'virgin'. Gendered ideas of the landscape as feminine ran through the marketing of the New South Wales Government Railways brochures. With or without leaving the comforts of the train, according to the New South Wales Government Railway brochures, the gendered language suggested that men who travelled by train along this coastal route could now join the ranks of previous male explorers in the pleasures of the pioneering work of discovering and experiencing wilderness. Equally, circulating the European myth of wilderness among day-trippers, picnickers and hikers from Sydney helped to confirm the legal concept of *terra nullius* and silence the uncompleted shift of land from Aboriginal custody into the ownership of colonisers.

Likewise, motor touring worked in similar and different ways from Britain. For affluent men and women the motor tour in the early 1900s became a mechanism of class distinction. Following the trend in Britain at this time, the motor tour was about stops, scenic spectacle and social status rather than speed. In the Illawarra, the ocean, beach, escarpment and rainforest became a tourist resource, part of the wonder in which motor tourists could experience 'The Garden of New South Wales'. For Tourist Associations, publication of automobile guidebooks enabled the recirculation of foundational stories about the Illawarra as an earthly paradise. Again, the European myth of earthly paradise helped legitimise colonial ownership over the land through silencing an Aboriginal presence. With the desire to experience the Illawarra as an earthly paradise, the motor tourist was disinclined to question the myths of *terra nullius* on which British occupation of Australia depended. The Illawarra was historised in British terms and endowed with British meanings. While the motor tourist had control over the speed, distance and direction of travel, guidebooks encouraged the motor tourist to follow particular routes, stop at designated places and inscribe particular meanings onto places. The Illawarra was cast as picturesque. Witnessing picturesque views of the ocean also became one way to generate narratives about how the motor tourists were aligned to the project of nation building and forging a British Australian identity.

4 The 'Brighton of Australia' Becomes the 'Sheffield of the South': Knowledge, Power and the Production of an 'Industrial Heartland' in an 'Earthly Paradise'

People are speaking in glowing terms on the manner of the [surf] club members, who are doing work voluntarily to make our beach [North Beach Wollongong] one of the best ... Each day they [members of the surf club] are working on the drain, and the work is very hard. The Council now sees how enthusiastic our members are trying to beautify our beach. It is up to them [Council] to give us assistance
Illawarra Mercury, 1931b

In December 1931, publicity officer, 'Sunshine', of the North Beach Surf Club sent this article to the *Illawarra Mercury* seeking Council recognition and support of the voluntary labour of surf club members in building a new drain. Seemingly, the voluntary work that was regarded as improving North Beach Wollongong through improving amenities for visitors to the beach was invisible to Council. This article was just one of many written to the *Illawarra Mercury* in the first four decades of the 20th century communicating the frustration among the surf clubs and Tourist Associations at the aldermen's lack of concern for beach cultures. How the space of the beach affords leisure and tourist potential was seemingly not valued by Councils. Even at the height of the surf lifesaving movement in Australia in the 1930s, Wollongong Council was apparently resistant to fund even the most basic of repairs.

This chapter draws on Michel Foucault's (1991) notion of 'governmentality' and 'regime of truth', discussed in Chapter 2, to help understand how discursive structures are implicated in the judgement of aldermen's decisions of what were appropriate practices at the beach. On the one hand, the reluctance of aldermen to acknowledge the work of surf clubs, or provide financial support to the material infrastructure of beach cultures, maybe in part explained by how 'truths' for aldermen about the beach were apparently established through Judeo-Christian ethics of the body. As already discussed in Chapter 2, what aldermen apparently knew about the pleasures of the beach as a sexualised space brought the tourist as surf-bather into being on the beach only as 'erotic', 'unmanly', or as a 'larrikin'. And, as discussed in Chapter 3, the arrival of the surf-bathers by train from Sydney brought new regulations and practices of the bathing ordinances and the new spaces of the bathing reserve.

On the other hand, the judgement of aldermen about tourism and beach culture was also framed through economic discourses of modernisation. Orientating the Illawarra through economic discourses of modernisation helped to stabilise the spatial boundaries that categorised the region as a centre of the coal and steel industries. Beaches were forged as sites where bodies came to work in the export of coal, rather than visit as a tourist. In this sense, economic discourses of modernisation as a set of coherent words, actions, institutions and infrastructure, produced a new regime of truth about the Illawarra as an Australian 'industrial heartland' rather than an 'earthly paradise'. The beach was not external to this process, but reconstituted through it as a place central to the shipping of coal.

To explore the forging of the Illawarra through economic discourses of modernisation this chapter is divided into three sections. The first section pays attention to particular people with the authority to speak about what were appropriate economic activities. Attention then turns to how normative discourses surrounding 'progress' through economic investment in resource and manufacturing sectors informed decisions of aldermen about what were appropriate investments for the region. Aldermen ascribed to economic modernisation discourses that aligned progress with industrialisation, increasing exports, productivity and profit. At the same time, economic modernisation also aligned increased manufacturing productivity with well-being and civilisation. Discourses of economic modernisation brought the 'category' of the Illawarra economy into being for aldermen and the State government through the statistics of manufacturing production, revenues and exports.

The second section explores how economic modernisation discourses and the production of the Illawarra as a 'manufacturing heartland' is implicated in the judgement by aldermen and businessmen not to support investment into the material infrastructure of beach cultures. Decisions taken among the aldermen and businessmen limited the field of possibilities for the Illawarra to ever remain a beach resort. The aldermen never welcomed tourists as consumers to the Illawarra and positioned surf-bathing not only as indecent but also as a 'craze'. The unwelcoming climate that the tourism industry encountered in the Illawarra, including the rejection of requests to provide entertainment facilities on the beach such as camel rides and merry-go-rounds, worked against investment in accommodation. The colonial regime of truth of the Illawarra as the 'Brighton of Australia' still voiced by Tourist Associations, the New South Wales Government Railways, bank managers and the car industry was joined by a new regime of truth of the Illawarra as the 'Sheffield of the South' and an Australian industrial heartland.

The final section examines the implications of subsequent amendments to the bathing ordinances that had brought into existence the bathing reserve. Councils became legally responsible to support the club house facilities, equipment and activities of surf clubs and surf lifesavers. Surf-bathing became an item on Council budgets. By claiming coal and steel as a prized industry and future vision for the regional economy during the 1920s and 1930s, Illawarra aldermen closely monitored all infrastructure requests and expenditure related to improving or maintaining beach facilities by the Surf Life Saving Association. Indeed, Wollongong Council erected a barbed-wire fence at North Beach to restrict access to only paying surf-bathers.

Illawarra Geographies of Economic Modernisation

The composition of councils was a crucial starting point to investigate who authorised investment in the coal and steel industry rather than tourism. Between 1900 and 1940, Peter Sheldon (1997) and Jim Hagan (1997) noted that prominent middle-class British-Australian business*men* dominated councils. Sheldon (1997: 102) argued that aldermen consisted of managers from 'manufacturing, collieries and commerce', the same groups that dominated the electoral and lobbying processes for the New South Wales Legislative Assembly. Therefore, despite a specific 1918 Act which gave women the right to take elected local government office, it was the aldermen, and more specifically businessmen, who had considerable vested interest over the economic agendas of councils.

In the early 1900s, for Illawarra aldermen, the 'economy' was created through the statistics of highly profitable Illawarra coal mining, exporting and processing industries. Coal and coke production peaked in 1902 with 959,134 tons of coal, and 51,592 tons of coke shipped from Wollongong (Lee, 1997a: 52). Economic modernisation discourses around these production figures did not simply describe the Illawarra economy for the aldermen, but helped create the economy, sustaining the coal and steel industries as the dominant form of economic power and maintaining ideas of progress. Therefore, far from the regimes of truth that generated anticipations, experiences and orientations towards the Illawarra as 'the Brighton of Australia', the beaches and escarpment became entangled in discourses of economic modernisation that bestowed progress as manufacturing and mining activities (Figure 4.1).

Economic modernisation was the foundation stone for State and municipal government policy, and pivotal to economic progress were manufacturing industries. At Federation (1901) it was taken for granted that manufacturing activities were imperative to national security. New economic modernisation discourses appeared central to informing aldermen's judgement of what was counted as acceptable industrial practices within the Illawarra. Consequently, in the first few decades of the 20th century the Illawarra was a focus for State industrial policies centred on import substitution and heavy industry (Sheldon, 1997). In 1900, for example, work

Figure 4.1 'Brighton Beach, Wollongong' (*Source*: Unknown. From the collections of the Wollongong City Library and the Illawarra Historical Society). Coal from Mount Kembla Colliery waiting shipment from Wollongong Harbour (Belmore Basin)

commenced in Port Kembla on an improved harbour at a site chosen by the State government for coal export and heavy industry (Lee, 1997a).

In 1908, the radical restructuring of economy required by the accountability of truth to manufacturing industries was underscored at Port Kembla by the opening of the Electrolytic Refining and Smelting Company (ERS, a subsidiary of the Mount Morgan Gold Mining Company). Glenn Mitchell (1997) discussed how Wollongong Council was fully supportive of the request by the ERS to build a refinery near the newly built harbour. The ERS indicated the refining and smelting of copper ore would produce sulphur fumes which 'may affect vegetation in the vicinity of the Works' (Mitchell, 1997: 146). Wollongong Council, however, was not to be deterred. Indeed, it appeared that the Mayor of Wollongong explained that sulphur dioxide would act as a disinfectant and help increase milk yield in the local cows (Mitchell, 1997). As the work of Foucault suggested, discourses delineate, produce and reinforce relations between what it is possible to think, say and do. Within the discourses of economic modernisation new 'truths' of sulphur dioxide became possible; the Mayor mistakenly constituted sulphur dioxide as a disinfectant. In March of the same year, in the context of Kiama Council funding the building of dressing sheds at Kendall's Beach, Alderman Hindmarsh (Mayor of Kiama) is reported in the *Kiama Independent* as dismissing surf-bathing among young people as a trend:

> I am not in favour of the object [dressing sheds] ... The Council should not encourage what I think is only a craze, and if further indulged in there is almost certain to be an accident – several fatal accidents have already occurred at Kendall's Beach. (*Kiama Independent*, 1908)

In the judgement of Mayor Hindmarsh, surf-bathing was a momentary fashion. Discussing surf bathing as 'a craze', he attempted to undermine arguments about the capital expenditure required to construct bathing resort facilities.

Instead, following the opening of the ERS, Wollongong Council approved a cluster of manufacturing activity (Sheldon, 1997). Discourses of economic modernisation were again implicated in the approval of a state-owned power station at Port Kembla, opened in 1914. According to Henry Lee (1997b: 63) the power station was 'expanded progressively to accommodate the requirements of new industries, municipalities and State government authorities'. The onset of World War I and the discursive structures of nationalism only helped to reconfigure Wollongong as an industrial heartland of Australia. By the end of the 1920s, Wollongong had a cluster of major industries at Port Kembla. Along with ERS, Metal Manufactures Limited (MM) manufactured copper bars from ERS into wire, cable, sheets, tubing and alloys (instruments of war); Australian Iron and Steel (AIS) constructed a blast furnace in 1927 and; Australian Fertilisers Ltd (AFL) turned waste sulphuric acid from ERS into superphosphate (Kelly, 1997). As Mitchell noted in discussing the opening of manufacturing plants in the early 20th century:

> These enterprises, along with a developing harbour and prosperous coal mines along the escarpment, realised the dreams of those local government officials and business people who equated progress with industrial development ... Those businessmen, who in the nineteenth century had called for the creation of a 'Sheffield in the South' or a 'new Birmingham', now had a substantial part of their dreams realised. (Mitchell, 1997: 147)

According to the regime of truth of economic modernisation that forged the Illawarra as an Australian manufacturing centre in the early 1900s, there was no pollution only job security, profit and national security (Figure 4.2).

In 1935, with the onset of the Great Depression, AIS merged with Broken Hill Associates. The merger gave Broken Hill Proprieties (BHP) the monopoly of iron and steel production in Australia. In 1937, with the purchase of

Figure 4.2 Australian Iron and Steel Pty. Ltd., Port Kembla (*Source*: Unknown. From the collections of the Wollongong City Library and the Illawarra Historical Society)

Mt Keira and Bulli mines BHP became the largest employer in the district and ultimately, 'Australia's largest industrial organisation for much of the twentieth century' (Fagan & Webber, 1999: 112). BHP continued to determine most of the employment and economic growth in the region in the late 1930s and 1940s (Lee, 1997b). By 1940, the Illawarra was one of the fastest-growing locations for manufacturing employment and industrial-related economic growth in Australia (Castle, 1997). As one of Australia's 'industrial heartlands', the Illawarra became a region imagined as inhabited by particular people (often coded by class), and associated with particular kinds of manufacturing and mining practices. Together these did not resonate strongly with how tourist associations and related industries had created a regime of truth that framed the Illawarra as a tourism destination – including earthly paradise, picturesque and escape.

Economic Modernisation Discourses and the Beach

Between 1900 and 1940 it appears that many aldermen equated fumes and smoke plumes, rather than surf-bathers with economic and material progress. Hence, during the first decades of the 20th century, the aldermen invested both financially and rhetorically, in economic modernisation discourses that brought the Illawarra into existence as a 'manufacturing heartland'. This new category of 'the Sheffield of the South' left little opportunity to appreciate investment as acceptable or appropriate in understanding the Illawarra as the 'Brighton of Australia'.

Advocates for the tourism industry repeatedly presented councils and businesses with the economic statistics that helped bring the category of 'tourist traffic' into being. However, investors and aldermen were seemingly unwilling to embrace tourism. In an article in the *Illawarra Mercury*, February 1911, an anonymous advocate for the tourism industry asked the question: 'Do Tourists Help Wollongong?' The article points out how economic modernisation discourses had made the category of 'tourist traffic' invisible:

> It was upon the question of catering for tourists and residents in the matter of increased bathing facilities, however, that the most extraordinary statements were made. When aldermen can be found who will publicly state that they have no faith in the value of the tourist traffic, it is difficult to take them seriously ... Of course Thirroul and Austinmere (sic) are being rapidly built up into thriving centres exclusively by tourists, but we are apt to overlook the fact. (*Illawarra Mercury*, 1911)

The suggestion was that tourists were invisible to aldermen, despite the increasing number of holiday homes in Thirroul and Austinmer. As the newspaper article went on to point out, without acknowledging the economic returns of tourism there were little possibilities for new investment and new tourist facilities:

> It must be admitted that Wollongong cannot enjoy the full benefit of the tourist traffic or expand with it as it should, unless our landowners can be induced to make the land available either by selling it or by erecting buildings. (*Illawarra Mercury*, 1911)

The point made by this anonymous author demonstrates how tourism sat outside the rules of economic discourse of aldermen and landowners.

Over the early decades of the 20th century the Tourist Associations also lamented the lack of support for the tourism industry from local government and businesses. For example, in January 1935 the Wollongong Tourist Association noted the lack of Council interest in maintaining the sets of ideas that made the Illawarra District visible as a tourism destination:

> At the present time very little attention is being devoted to boosting the attractions for tourists in the district... The Wollongong Tourist Association is composed of a small band of workers who at present are as a voice crying in the wilderness, owing to lack of public support... We realise that without the co-operation of the local governing bodies, success could not be assured, and are, therefore, asking your council to become identified with the movement by appointing official representatives to attend a meeting at 8 p.m. on Friday 13, 1935... The Illawarra District has attractions superior to all the centres named, but no effort is being made to make them widely known. (*Illawarra Mercury*, 1935b)

Tourism advocates lacked the political authority and societal apparatus of municipal authorities in their circulation and reproduction of 'truths' about tourists. At a Wollongong Tourist Association meeting in June 1935, the President of the Association, Mr Osborne, stated that during his presidency of the last seven years 'not one businessman had been on the committee'. He continued by saying:

> that unless the business people of the town woke up to a sense of their responsibilities, it would be useless to organise for the purpose of attracting tourists. (*Illawarra Mercury*, 1935c)

In the mid-1930s both municipal authorities and business people were seemingly disinterested in assisting tourism. According to the Wollongong Tourist Association, business people overlooked how the region remained visible as a tourist destination and the economic potential of promoting beach cultures.

Equally, those entrepreneurs who wished to provide entertainment facilities for tourists faced opposition from Council. How aldermen regulated tourism was illustrated in fortnightly decisions of Council meetings published in the *Illawarra Mercury* concerning requests to promote entertainment and events. Requests designed to enhance the pleasures of beach resorts came from individuals, as well as organisations such as surf clubs.

In the 1920s, the annual surf carnival was a popular mass tourism event, with sometimes thousands of people in attendance (Figure 4.3). Permission was required from Council to organise a surf carnival. For example, in 1929: 'Application is also made for the permission to conduct the annual surf carnival on Austinmer beach on Saturday, the 28th December. Granted' (*Illawarra Mercury*, 1929c). And in 1937: 'From Thirroul Surf Club asking permission to hold a surf carnival on 6th March – Granted' (*Illawarra Mercury*, 1937f). However, not all requests for carnivals were approved. In 1930, Corrimal Surf Club was refused permission to hold their carnival on Anniversary Day Holiday because of a ruling that 'no carnivals be allowed

Figure 4.3 Surf Carnival, Woonona Beach (c. 1910) (*Source*: Unknown. From the collections of the Wollongong City Library and the Illawarra Historical Society)

on a holiday' (*Illawarra Mercury*, 1930a). Likewise, until the mid-1930s no carnivals were allowed to be held on a Sunday. Drawing on Judeo-Christian regimes of truth some aldermen understood the pleasures of a surf carnival to be out of place on the beach on a Sunday.

Judeo-Christian regimes of truth surrounding pleasure perhaps also operated to ban the provision of entertainment facilities on the beach. In 1929, for example, there was a request from C.H. Phillips of Sydney who applied to Shellharbour Council for permission to 'erect and run various stalls at the Lake Entrance during the holidays' (*Illawarra Mercury*, 1929d). The *Illawarra Mercury* recorded that 'permission was refused'. At the same meeting, applications were also refused for 'merry-go-rounds and side shows'. In 1936, an application for an amusement ride was rejected, along with an application for camel rides on the beach:

> Application is made by H. J. Offer for permission to conduct camel rides on beaches during the coming holiday season. Recommended that permission be refused. (*Illawarra Mercury*, 1936b)

Clearly, the majority of aldermen did not support commercial beach amusements. Aspects of beach culture such as band recitals, black and white minstrel shows, Punch and Judy shows, donkey rides and stalls were commonplace elsewhere (Huntsman, 2001). In contrast to the willingness aldermen demonstrated to facilitate investment into manufacturing industries and coal mining, investment into tourism facilities were banned or stalled. As previously discussed in Chapter 3, perhaps the reluctance to approve commercial amusements is tied to aldermen claiming that the virtues of the respectable middle-class configured 'mass' tourism as vulgar and a hullabaloo.

The Bank of New South Wales records (now known as Westpac) confirmed the paradox of this region visible to visitors as a beach resort, but lacking investment in tourism infrastructure. Throughout the 1920s and 1930s, many bank managers categorised the Illawarra as a 'seaside resort'. For example, one bank manager reporting on the building of the cross-country railway between Mossvale and Port Kembla in the early 1930s understood the branch-line as an appropriate action because it would enhance the flow of visitors to Wollongong as a 'sea-side resort':

> The Mossvale railway, which is expected to be finished in about 12 months, should open up as avenue to draw visitors from the west to Wollongong as a sea-side resort. (Wollongong Branch, 1930. Inspector's Report Metropolitan Division New South Wales: 1909–1931)

Similarly, one bank inspector writing in 1934 was optimistic in his categorising of Wollongong as a 'tourist resort':

> General outlook is promising . . . Wollongong is also favourably situated for a Tourist Resort especially during the summer months. (Wollongong Branch, 1934. Inspector's Report, Bank of New South Wales Central Division: 1932–1940)

Yet, at the same time the bank managers and inspectors constantly reported the lack of investment throughout the 1930s into tourism accommodation. The bank records suggest that economic investment fell generally during the Depression of the 1930s: 'progress at a standstill, business community is having a lean time' (Wollongong Branch, 1930. Inspector's Report Metropolitan Division New South Wales: 1909–1931); 'building operations very quiet' (Port Kembla Branch, 1930. Inspector's Report Metropolitan Division New South Wales: 1909–1931). However, the lack of investment into accommodation pre-dates the Depression given that in February 1930 the Wollongong Tourist Association referred to a 'shortage of accommodation for tourists in the district' (*Illawarra Mercury*, 1930b) and the Austinmer Surf Club reported that 'hundreds of people have been unable to get accommodation' (*Illawarra Mercury*, 1930c). As discussed in Chapter 3 the assemblage of State and private industries that facilitated rail and car mobility continued to make the Illawarra visible through the discourses of an earthly paradise.

According to the bank records a shortage of accommodation continued after the Depression. In 1935, for example, the Kiama bank manager reported that:

> There is a strong demand for all classes of real estate . . . [Kiama is] regaining its popularity as a holiday resort but lack of suitable accommodation is a drawback. (Kiama Branch, 1935. Inspector's Report, Bank of New South Wales Central Division: 1932–1940)

Mr Cahill of the Wollongong Tourist Association confirmed this shortfall at a September 1935 meeting. He is reported to have said that: 'The trouble to-day is not to get tourists . . . but to find accommodation for the visitors' (*Illawarra Mercury*, 1935a). In 1936, 1937 and 1939, bank manager reports in Wollongong simply stated in the context of tourism accommodation that: 'There is an acute shortage of dwellings' (Wollongong Branch, 1936, 1937, 1939. Inspector's Report, Bank of New South Wales Central Division: 1932–1940). In 1937, further evidence of a lack of entrepreneurial capital flowing into accommodation was found in an anonymous article in the *Illawarra Mercury* concerning 'The Grand Hotel':

> People always seemed to be inquiring for hotel accommodation in Wollongong, and unable to get it. (*Illawarra Mercury*, 1937g)

Indeed, in the 1930s, the lack of hotel accommodation was only one dimension of the lack of apparent willingness to invest in facilities to create leisure and tourism spaces. With the important exception of the opening of the Continental Baths in 1928 (see Chapter 6), funded because of the increased popularity of swimming, Wollongong could not boast of any 'seaside' resort facilities. At this time, for example, Coogee, in Sydney, possessed the Palace Aquarium, with associated swimming baths and amusement grounds, while Bondi, in Sydney, was replete with an Aquarium, and 'Wonderland City' – 'an amusement complex containing sideshows ... slippery dips and underground rivers' (Huntsman, 2001: 40). The next section turns to further examples that illustrate how the intersection of discourses of sexuality and economic modernisation shaped aldermen's judgement on the practices thought of as suitable within the bathing reserve in the 1920s and 1930s. At this time, the Council's decision appeared to prevent certain popular resort-style activities.

Bathing Reserves and Council Budgets

As discussed in Chapters 1 and 2, discourses of sexuality helped give rise to the spatial boundaries of the bathing reserve. However, in 1919, amendments were made to Ordinance No. 52. Under the 1919 Local Government Act municipal authorities continued to be responsible for governing public baths, bathing costumes and dressing pavilions. In addition, councils became responsible under Clause 1. 'Baths and Bathing Facilities' to provide, control and manage:

> club and drill rooms, appliances and materials for life-saving clubs; ...works and appliances for the protection of bathers from injury, drowning or sharks and; life-savers and life-saving or swimming instructors. (Clause 1, Ordinance No. 52, Local Government Act 1919)

By-laws put the surf-bather at centre stage of Council budgets. Suddenly, a whole range of ongoing infrastructure and personnel costs were required for the protection of surf-bathers, a subject only 10 years earlier Wollongong aldermen had discursively brought into being as unmanly. Wollongong Council did not initially respond favourably to sets of ideas that constituted the subject of surf-bathers in terms of expenditure in Council spread sheets. For instance, in 1929, the spatial boundaries of the bathing reserve at North Beach were materialised by the use of a barbed-wire fence and the implementation of payment for beach access. In December 1929, Wollongong Council outlined the reason behind the boundary marker:

> With regard to the fence on the North Beach, it was being placed so that persons who come down to the beach in costume will be forced, more or less, to pay the same as those who use the sheds, to help defray the cost of the life saver. (*Illawarra Mercury*, 1929e)

Access to the lifesaver at North Beach was restricted to only those who payed. The fenced enclosure that appeared on North Beach was how State by-laws had brought the surf-bather into existence as a major expense in Council budgets. Unlike the dressing shed which appeared in bathing reserves to make private the sexual body, the fence on North Beach speaks to the discourse of economics.

Tourism advocates and beach users protested at the idea of paying for the pleasure to bathe in what was taken-for-granted as a common property resource; the sea. On 20th December 1929, the *Illawarra Mercury* reported the outcomes of a meeting held by the Wollongong Branch of the South Coast Tourist Association:

> A long discussion took place in regard to the action of the Wollongong Council in fencing portion of the North Beach surfing area. It was contended by several speakers that this would have a detrimental effect so far as tourists were concerned, and would nullify the efforts of the branch to attract tourists to the town. It was decided to invite all public bodies in the town to send delegates to a monster deputation to the Council in reference to the matter. (*Illawarra Mercury*, 1929f)

For the Wollongong Tourist Association, the concept of user-pay beach access worked against the appeal of Wollongong as a resort. The Tourist Association sent a letter of protest to Wollongong Council and rallied in protest at the beach. An anonymous article in the *Illawarra Mercury* on the 20th December reported that that the fence was demolished, apparently by its opponents:

> A wire fence had been erected from the end of the men's sheds to the cutting and on Wednesday morning it was found to have been pulled down. (*Illawarra Mercury*, 1929g)

The construction of the fence suggested that Wollongong Council constituted surf-bathers primarily as cost and had little understanding of how the spending of tourists operated within the economy. Just over a year later, a Parks and Baths Committee Report recommended that the: 'Barbed wire fence on the Nth beach to be removed' (*Illawarra Mercury*, 1931c). Surf-bathers were not going to be forced to swim in-between the barbed-wire fence because of how they were invented as an expenditure through Council budgetary practices. Nevertheless, Wollongong Council was less than willing to support surf-bathing through funding surf clubs and lifesavers, despite the by-laws.

The surf club records, for example, give numerous accounts of councils declining to fund essential lifesaving equipment and paid permanent lifesavers during weekdays. In October 1929, Coledale Surf Club asked Bulli Shire Council for a 'new surf reel, line and belt' but were told that the 'Fund is overdrawn' (*Illawarra Mercury*, 1929h). The following month, Woonona Surf Club requested Council to 'return' a surf reel given no replacement had been forthcoming (*Illawarra Mercury*, 1929c). In 1931 Thirroul Surf Club requested Council 'to again consider providing a reel and belt for the use of the Southern end of Thirroul beach – not to be supplied' (*Illawarra Mercury*, 1931d). Further, in 1938, a lifesaving reel was reported to be in an inoperable state on Bellambi beach:

> From Surf Life Saving Assn., Illawarra Branch, asking Council to remove the life saving reel from Bellambi Beach, as it is not in a fit condition for use. (*Illawarra Mercury*, 1938b)

Equally, discussions over paid lifesavers were frequently deferred in the early decades of the 20th century. In 1915 Bulli Shire Council was asked to employ a lifesaver 'for one month for duty on Austinmer beach about Christmas time'. However, the surf club was 'to be informed Council had no funds for purpose' (*Illawarra Mercury*, 1915). Again some 16 years later, in 1931 the *Illawarra Mercury* reported that: 'The whole position regarding the [paid] life saver at North Beach will be discussed by the Council prior to next swimming season' (*Illawarra Mercury*, 1931c).

Throughout the 1930s, deferred spending on bathing infrastructure remained apparent. For instance, in 1935 the *Illawarra Mercury* reported that some ocean baths were considered unsafe and needed urgent repair: 'walls around the Scarborough baths are not high enough to stop children being washed out'. Also, at Wombarra, 'a new door is urgently required on the baths' (*Illawarra Mercury*, 1935d). Again, in 1936 a shark-proof enclosure at Brighton Beach was not considered urgent at a Wollongong Council meeting:

> that Council consider a shark-proof net off Brighton Beach, Council will give the matter consideration. The matter was deferred. (*Illawarra Mercury*, 1936b)

These examples suggest that councils were either unwilling or unable to fulfil their duty-of-care by carrying out maintenance and safety responsibilities. Spending of limited Council funds on bathing practices did not appear to be a Council priority, despite legal requirements. Whatever the case, lack of investment into bathing infrastructure became one way to regulate ocean bathing practices through sustaining an understanding of surf-bathing in the Illawarra as unsafe.

Similarly, the indifference to the surf lifesaving movement by Wollongong Council was apparent in the fraught relationship with surf clubs. For example, in 1935 Wollongong Surf Club reported on the difficulty of convincing aldermen of the value of investing in surf sheds:

> After a considerable number of years trying to persuade the various aldermen of the Wollongong Council that we were deserving of new surf sheds .... (*Illawarra Mercury*, 1935e)

Again, in 1935, Woonona Surf Club was reported to 'continue negotiations' with Bulli Shire Council in order to obtain a new club house (*Illawarra Mercury*, 1935f). And, in 1936, Mr Sheldon, the Wollongong Surf Life Saving Club president, stated at a foundation stone ceremony of a new surf pavilion at South Beach that: 'after many years of hard battling the new building was now a reality' (*South Coast Times*, 1936). Members of the surf club and some dignitaries understood the surf pavilion within the discourses of progress. Mr W. Davies (M.L.A.) (Member of the Legislative Assembly) at the opening of the surf pavilion, described Wollongong as 'a town that was forging ahead more rapidly than any other in the Commonwealth, and one which had potentialities greater than any other in the Commonwealth' (*South Coast Times*, 1936). While the President of the Surf Life Saving Association of Australia, Mr Adrian Curlewis, is reported to have said that the surf club house was outstanding:

> He had visited every surf dressing pavilion in the metropolitan area, but had to come to Wollongong to see the best in the State. (*Illawarra Mercury*, 1936c)

However, for the Wollongong Surf Club the 'battle' continued with Council long after the official opening of the new £6000 surf pavilion. In this case, funding from Wollongong Council was a type of loan and the debt became a source of controversy. Mayor Kelly had promised that 'Council would be behind any organisation which had for its objective any move for the progress of the town' (*South Coast Times*, 1936). Consequently, the Wollongong Surf Club members repeatedly argued that 'the responsibility to pay off the... building cost should rest with the Wollongong Council' (*Illawarra Mercury*, 1936c). However, Wollongong Council refused to accept this argument. Surf pavilions were not integral to understandings of progress framed by economic modernisation. Exemplifying how Wollongong Council's decision-making was framed by economic modernisation was a proposal in 1937 to construct a sewerage outlet at North Beach. In 1937, Wollongong Surf Club stated that:

> It has been rumoured that the Council intend to extend the sewerage system across our surf. We hope this is not right, and we strongly protest against any such scheme. (*Illawarra Mercury*, 1937c)

Newspaper and surf lifesaving records suggest that neither surf lifesavers nor surf-bathers were valued by Wollongong Council in the 1930s despite the Surf Life Saving Association of Australia narratives of a nation forged on the beach.

Throughout the 1930s, letters published in the media written by surf club members pointed towards how in comparison to Manly and Bondi, surf lifesavers were not valued by councils in the Illawarra. For example, Mr George Isedale, vice president of the North Wollongong Surf Club, is reported to have said that:

> records were being broken in Wollongong, as this was the only place where surf clubs were building for a council. However, this was the only way the clubs could get what they wanted and now the clubs had to pay back the borrowed money. (*South Coast Times*, 1936)

Conclusion

Stories in the public media about Council engagement with tourism suggested that in spite of the increased number of people arriving as tourists in the Illawarra, most aldermen were opposed to a resort style seaside tourism industry. This was despite acknowledgment among bank managers and bank inspectors who positioned the Illawarra as a 'seaside resort' and appeals by Tourist Associations and members of the surf lifesaving organisation to invest in tourism infrastructure.

Newspaper articles and council records are full of reports about the efforts of aldermen to prevent the pleasure economies of the seaside resort. Surf-bathing was not only positioned by some aldermen as a 'craze' but as an economic burden on Council. Evidence of aldermen's resistance to financing resort tourism is found in a reluctance to allow seaside-style entertainment, the establishment of a user-pays lifeguard system and failure to fund lifesavers' equipment. Perhaps by failing to fulfil or comply with legal maintenance and health and safety requirements, councils were hoping that the dangers of bathing would stop the growth in visitor numbers. Whatever the explanation, capital investment into resort infrastructure was never forthcoming. Instead, for the middle-class aldermen, drawing on discourses of economic modernisation, the regional economic future was tied to the coal and steel industries. Aldermen sanctioned State and private investment into a steelworks, copper smelter, electricity plant and the Illawarra coal mining, exporting and processing industries. They spoke with relish of the progress they associated with the manufacturing industry. Wollongong was reconfigured as the Sheffield of the South, and the Illawarra became an Australian industrial heartland. Tourist Associations continued to lobby councils, however the Illawarra became increasingly known as an industrial region.

5 'Battle for Honours': Surf Lifesavers, Masculinity, Performativity and Spatiality

... when Australia needs them [surf lifesavers], as some day no doubt she will, these men, trained athletes, tanned with the sun on the beaches, strong and brawny with the buffeting in the surf, will be well fitted to take up their trust and do duty for their country
Sydney Morning Herald (1908: 7)

Bronzed surf lifesavers since the 1930s became a symbol of the Australian nation. Bronzed surf lifesavers helped to create the illusion that Australia was a natural, timeless entity. Yet, as A.W. Relph, founder of the Manly Surf Club, writing in the *Sydney Morning Herald* in September 1908 reminded us, neither the nation nor national identity was natural, but socially constructed. As Benedict Anderson (1983: 6) argued they are 'imagined'. Australian national identity was in part built on stories about surf lifesavers and surf-bathers as 'good' citizens. For example, Relph's words exemplified how surf lifesavers offered one way of being a good Australian citizen by becoming bronzed, fit, strong and well-muscled. In turn, this enabled surf lifesavers during World War I and World War II to fulfil an obligation of citizenship; the defence of the Australian nation.

In the first decade of the 20th century, some Sydney authorities sought to discipline the bodies of surf-bathers through regulating men's bathing costumes. Colonial bourgeois norms that defined masculinity in terms of dignity translated into laws that restricted daylight surf-bathing and constituted the surf-bather within the category of moral degenerate. In 1907, the regulation of bathing costumes for men sparked a backlash against municipal authorities by surf club members in Sydney. Councils were criticised for failing to appreciate how the bodies of surf lifesavers were imbued with the same mythical qualities of bushrangers, drovers and diggers on the goldfields. Men who lived in the open air, learnt to 'shoot' waves, were also resourceful, independent and shared the social bonds of mateship. Surf-bathers fitted into the Australian nation's self-construction through the Romantic mythologies of the 'bushman'. The Surf Bathing Association of New South Wales was one outcome of the protest by surf-bathers. In 1908, alongside improving facilities for surf-bathing, the objectives of this Association were to enforce the bathing ordinances outlined in Chapter 2. On the one hand, the discourses circulated by members of the Surf Bathing Association of New South Wales helped reconfigure the subjectivities of those individuals within the boundaries of the bathing reserve. On the other hand, the discourse of the surf lifesaving movement helped to stabilise sexed, gendered and raced ideas of the Australian nation. As argued by Douglas Booth (2001), John Ramsland (2000), Grant Rodwell (1999) and Cameron White (2007), how the surf lifesaving movement redefined Australian citizenship did not extend rights to all members of the nation state. These authors argued that the identity of the surf lifesaver excluded from the full benefits of citizenship other social categories such as Aboriginal Australians, women and homosexuals.

This chapter draws on the ideas of Judith Butler (1990) to explore the lifesaver as an effect of discourse. This chapter examines the exercise of discourses of masculinity of the Surf Life Saving Association (SLSA) in the Illawarra in the first decades of the 20th century. At the beach, the relationship between the bodies of surf-bathers and lifesavers and the gaze of others was legitimated as an appropriate display of Australian white heteromasculinity. The rituals and cultural traditions of surf lifesaving clubs stabilised frames and reading practices which protected the bodies of surf lifesavers dressed in bathing costumes from slipping into significations of the sexual or feminine. Through the coherence of the performative subject of the surf lifesaver their bodies became a privileged site where semi-nakedness was seemingly read in a non-sexual way through the relationships that comprised the surf beach. The SLSA created normative expectations about the surf beach as a masculine space and the ways that surf club members would do masculinity at the beach.

More specifically, the aim of the chapter is to explore whether the members of the SLSA in the Illawarra reiterated hegemonic discourses of lifesavers by celebrating lifesaving as heroic and masculine. The chapter primarily draws on surf club records from 1908 to 1940 to explore how men negotiated masculinity through membership of the Illawarra Branch of the SLSA. We remain mindful that the surf club records are partial. There is a clear gendered dimension. All surf club archival material was written by men, including reports on women's surf clubs and activities. There are no narratives in the SLSA archive from women. Further, the material provides insights into how primarily the membership wished to portray itself to a public audience. Missing are personal reflections and life histories that would reveal further the complexities and contradictions within the surf lifesaving movement (e.g. see Ed Jaggard's (1997) discussion of surf lifesavers as 'chameleons').

The chapter is divided into three main sections. The first section discusses the institutional context of the SLSA. As Butler argued:

> genealogy investigates the political stakes in designating as an *origin* and *cause* those identity categories that are in fact the *effects* of institutions, practices, discourses with multiple and diffuse points of origin (italics in original). (Butler, 1990: x–xi)

Is the SLSA valorised in the Illawarra? The next section examines how the discourses of the SLSA were implicated in the production of the concept of the Illawarra surf beach and the judgement of surf-bathing and lifesaving practices. Three main themes illustrate how the SLSA stabilised the surf beach as a heterosexualised space and the heteromasculine identities of lifesavers: the athletic; vigilance and service; and the family. The last section examines stories about tanning and surfing. Surf club members reiterated hegemonic discourses of masculinity through tanning and surfing. Yet, at the same time Illawarra surf club members who sun-bathed and surfed for pleasure unsettled and complicated what the SLSA deemed appropriate practices at the surf beach. The hedonism of tanning and surfing momentarily destabilised expected behaviours of members, causing 'lifesaver trouble'. Following arguments of both Jaggard (1997) and Booth (2001), some Illawarra surf club members could also be categorised as 'larrikins' through challenging social norms of the SLSA.

At the Surf Beach with the Illawarra Branch of the SLSA

Following Michel Foucault (1980), the SLSA can be conceived as an institution that constituted 'regimes of truth', power and subjectivity. The SLSA regimes of truth operated to produce geographical discourses that stabilised understandings of surf lifesavers as belonging on the beach. First, they are humanitarian, in that they ensure a politics of freedom and quality could operate at the surf beach. Second, they are Australian; surf lifesavers represented the 'natural' progress of Australian society. Third they are masculine; surf lifesavers were 'naturally' strong, bronzed and disciplined. Finally, surf lifesavers became taken as self-evident embodiments of an Australian white heteromasculinity.

Regimes of power associated with eugenics underpinned how surf lifesavers became self-evident embodiments of a modern, healthy, disciplined, virile and humanitarian Australia. As Rodwell (1999) argued, informed by eugenics, SLSA bodies on the beach became entangled in a plethora of regulatory schemes to monitor, shape and discipline surf lifesavers (and non-surf lifesavers). SLSA members were to move and behave in ways consistent with an ordered, regulated, and disciplined Australian society, attenuating the 'white heteromasculinity' of surf lifesaving. Eugenicist arguments resonated with those who ascribed to the misplaced 'truths' of 'racial decay'. As Christina Cogdell (2003: 39) pointed out, eugenicists simultaneously built upon and displaced the Darwinian idea of evolution, arguing that 'natural selection' had to be replaced with 'rational selection'. Fit, muscled and tanned surf lifesaver bodies at the beach could be interpellated as illustrative of rational selection at work. Consequently the tanned, fit, strong bodies of SLSA members could be appreciated in terms of their eugenic attributes, signifying a pure, virile, white and healthy race. For example, an article in the *Sydney Mail* emphasised tanning as a mechanism of social distinction, suggesting that 'a sun-browned skin always wins respect' (*Sydney Mail*, 1906c: 619). Similarly, a year later the *Evening News*, reported that:

> a well-browned skin is much healthier than a white one... So the sun-worshipper looks with pity upon his pallid brother as one who stupidly neglects a most evident good, and, in fact one who falls short somehow in the standard of true manliness. (*Evening News*, 1907, cited in Booth, 1991: 139)

The racialised, gendered and therapeutic truths of how the sun transformed bodies were not lost on executives of the SLSA. For instance, the *Sydney Morning Herald* reported on W. Tonge's and A.W. Relph's beliefs in the transformative qualities of tanning for British men:

> On a Saturday afternoon or a Sunday morning in the summer it is a sight worth looking at to see the hundreds of fine brown-skinned specimens of manhood on the beach. (*Sydney Morning Herald*, 1907c: 6)

Tanned skin became self-evident of gender differentiation and establishing hierarchies of social difference. On the beach, the confluence of race and gender through the tanned body of the surf-bather are integral to understanding the racialised form of Australian nationhood. Drawing on the fantasy of how 'Australia' came-into-being through surf-bathers, the context of gazing upon surf-bathing bodies was legitimised as one way to become intimate with the nation. For instance, in 1912, The New South Wales Surf Bathers Guide provided a useful example in which a surf-bathing body was legitimately gazed at by others, mediated by the convergence of racialised and gendered truths of tanning:

> One of the finest sights that can be witnessed is to see a dozen or more of the Brown Men of the beach on the crest of an incoming wave coming shorewards at a terrific pace. (*The New South Wales Surf Bathers Guide*, 1912, cited in White, 2003: 107)

The surf-beach became a privileged space where semi-nakedness took place, but the imperative to look was encouraged in order to show the Australian nation coming-into-being.

From January 1908, the geography of the Illawarra beaches was transformed by the progressive formation of surf clubs and the Surf Bathing Association of New South Wales (SBANSW) which became expanded to the Illawarra Branch of the Surf Life Saving Association of Australia (SLSAA) in 1929. In October 1907, the SBANSW was formed as a result of a meeting of members from a number of Sydney surf lifesaving clubs, the Royal Life Saving Society and the New South Wales Amateur Swimming Association. The first Illawarra surf club was established a year later and named the Wollongong Surf Bathing and Life Saving Club (refer to Table 5.1 for surf club association name changes between 1908 and 1929). During the early 1920s, Illawarra surf clubs became affiliated with the SLSA. Not all surf clubs were affiliated each year with the SLSA. Some surf clubs had no Bronze Medallion holders and it was a requirement for the club to have a Bronze Medallion holder to be affiliated. The Bronze Medallion was introduced in 1910 as the basic qualification to perform rescues. Bulli Surf Club, for example, did not have any members with their 'bronzes' until 1930. Other surf clubs simply failed to register on time. The later was the case for Thirroul and Bellambi Surf Clubs, in the 1934/1935 surf season (1934/1935 Illawarra Branch Annual Report).

Table 5.1 Name changes of Illawarra surf lifesaving associations

Name changes of Illawarra surf lifesaving associations
1929: Illawarra Branch of the SLSAA
1924: South Coast Branch of the SLSAA
1920: SLSA of New South Wales
1917: South Coast Surf Bathing Association
1908: New South Wales Surf Bathing and Life Saving Association

Source: Surf club records

Table 5.2 Surf clubs affiliated with the Illawarra branch of the SLSA between 1932/1933 and 1937/1938 surf seasons

1932/1933	*1934/1935*	*1935/1936*	*1937/1938*
Kiama	Kiama	Kiama	Kiama
Port Kembla	Port Kembla	Port Kembla	Port Kembla
Wollongong	Wollongong	Wollongong	Wollongong
North Wollongong	North Wollongong	North Wollongong	North Wollongong
Coniston	Coniston	–	–
Corrimal	Corrimal	Corrimal	Corrimal
Bellambi	–	–	–
Woonona	Woonona	Woonona	–
Bulli	Bulli	–	Bulli
South Thirroul	South Thirroul	South Thirroul	South Thirroul
Thirroul	–	Thirroul	Thirroul
Austinmer	Austinmer	Austinmer	Austinmer
Coledale	Coledale	Coledale	Coledale
Coalcliff	Coalcliff	Coalcliff	Coalcliff
Helensburgh/ Stanwell Park	Helensburgh/ Stanwell Park	Helensburgh/ Stanwell Park	Helensburgh/ Stanwell Park
			Shellharbour
			South Kembla
			Illawarra South

Source: Surf club records

Table 5.2 gives further examples of non-affiliated surf clubs between 1932/1933 and 1937/1938 surf seasons (indicated by –). The Illawarra Branch Annual Reports indicated how many clubs affiliated each season with the Illawarra Branch as well as the total of 'capitated members' in the Branch – that is, 'the Active members who are holders of the Bronze Medallion' (Table 5.3).

Appeals were regularly made for new members from the late 1920s. For example Mr Hanley, a surf lifesaving instructor, appealed to men to join surf clubs on the moral ground of self-sacrifice and civic duty in 1929:

> out of the thousands who enjoy the privilege of having a good time in the surf how many give serious thought to the position of the members of the clubs whose vigilance on the beaches makes it possible for them to have that good time in the surf in comparative safety – not many. Having

> this in mind, I would like to make an appeal to all eligible young men of the South Coast to join up with their respective clubs and make that club worthy of far better support from the citizens who enjoy the privilege of surfing safety . . . As surf clubs here badly need more strength, I hope this appeal will not be in vain, and that their members will be doubled. (*Illawarra Mercury*, 1929i)

Men of the Illawarra responded to such calls. By the late 1930s, there were 15 clubs affiliated with the Illawarra Branch. These clubs patrolled surf beaches from Helensburgh/Stanwell Park in the north to Kiama in the south (refer to Tables 5.2 and 5.3). The SLSA Illawarra Branch Annual Report (1940) noted that the Illawarra Branch was 'the largest Branch affiliated with the Surf Life Saving Association of Australia'.

Alongside the bathing ordinances, the SLSA became implicated in making visible the bathing reserve as a surf beach through a series of regulations. One way the surf beach became visible was to establish an opening and closing of the bathing season. Each year the SLSA announced the 'opening' of the surf beach in October. For example, in October 1937 a surf club publicity officer wrote: 'What a splendid day for the opening of the surf season' (*Illawarra Mercury*, 1937h). Similarly, each April, the surf beach was closed officially:

> This week end will, as far as surfing activities are concerned, bring the season to a close. (*Illawarra Mercury*, 1936d)

Announcement of a 'bathing season' brought the bathing reserve as a surf beach and the surf-bathing bodies under the surveillance, or disciplinary power, of lifesavers. The bodies of surf-bathers and lifesavers became entangled in discourses of the surf-beach-as-institution. Surf-bathers and surf lifesavers had to obey new sets of rules by performing their adherence to the SLSA codes of etiquette.

Table 5.3 The number of surf clubs and capitated members (bronze medallion holders) of the Illawarra branch between 1932/1933 and 1937/1938 surf seasons

Surf season	*Number of surf clubs*	*Capitation of branch*
1932/1933	15	352
1934/1935	13	424
1935/1936	12	448
1937/1938	15	443

Source: Surf club records

For example, during the bathing season, the beach became territorialised through the use of flags that marked the surf patrolled by lifesavers. The SLSA defined norms that governed the movement of surf-bathers through the use of flags. In the late 1930s, Wollongong Surf Club called attention to '[t]he importance of bathing between the flags' as 'most essential' (*Illawarra Mercury*, 1937i). Reminders from publicity officers to bathe only 'between the flags' appeared frequently in the *Illawarra Mercury*: 'Surfers are reminded that they must bathe between the flags on all beaches' (*Illawarra Mercury*, 1937j).

Similarly the bodies of lifesavers and the gaze upon them were disciplined through SLSA regulation bathing costumes. In December 1938, the *Illawarra Mercury* reported the curiosity surrounding the beach attire of surf lifesavers:

> Great interest is centred around the advisability of patrol members wearing some distinguishing colour either by caps or costume or both. (*Illawarra Mercury*, 1938c)

In 1939, distinguishing red and yellow caps and costumes became part of the standard SLSA patrol uniform. Through wearing particularly fashioned bathing costumes, surf clubs adhered to and upheld the SLSA order. Through wearing the regulation bathing costume, the idea of the near-naked body of the lifesaver was presented and viewed as naturally Australian and an asexual object.

Another way SLSA bodies self-imposed an order that made visible the surf beach and surf lifesaver was the bronze medallion programme; the basic qualification for surf lifesaving. Established in 1910, pioneer Manly Surf Club member, Arthur Lowe, spoke about the programme in the following terms:

> [H]e must obey. Fulfil all orders as to chores, swimming practices, drills, lectures and carry out everything necessary to enable him to pass one of the hardest examinations in the world ever set humans; that is, the test for the bronze medallion. (cited in Booth, 2001: 72)

As described by Arthur Lowe, the bronze medallion programme was a training ground in an acceptable male identity as defined by the SLSA. From the 1921 to 1922 surf season, according to Middleton and Figtree (1963), the Illawarra Branch had its own board of examiners to assure lifesavers met the standards expected of the role.

At one level the bronze medallion programme was illustrative of how eugenic discourses were channelled into designing drills and tests. Eugenic discourses became inscribed upon surf lifesaving bodies through training regimes and lessons on the 'correct' way to perform lifesaving. At another level the rules governing the award of bronze medallions to only men were implicated in how gender relations played out in and across the bodies of lifesaving bodies at the surf beach. The bronze medallion became a training ground for a particular style of heteromasculinity.

Although lifesaving and surf-bathing was enjoyed by men and women in the early 1900s, the bodies of women as surf lifesavers were not endorsed by the SLSA. Women in the Illawarra had formed their own surf clubs, gave displays of lifesaving skills and received Royal Life Saving bronze medallions (but not SLSA bronze medallions). However, for the SLSA, lifesaving was performative of a particular type of heteromasculinity in public space. Lifesaving enabled displays of a bravado heteromasculinity, given the body was in danger of injury, if not death, in the surf. For women to perform lifesaving would dissolve the SLSA's clear-cut notion of masculinity and femininity, and displace the heteronormative alignment of sex and gender. Through the restriction of the award of bronze medallions to only men, lifesaving was deemed more appropriate for men rather than women to display at the surf beach. The exclusion of women from the bronze medallion, and effectively surf rescue helped to territorialise and naturalise the surf beach as a place where only men should be involved in rescue, physical training and displaying their bodies. The exclusion of women was one way to constitute lifesaving as a masculine activity through the forging of homosocial bonds. Only in the 1980s were the normative gendered constructions of lifesaving and surf beaches challenged and women permitted to gain the coveted surf lifesaving Bronze Medallion.

Competitions between surf clubs were also organised to help enhance the heteromasculine qualities of surf lifesaving through opportunities to exhibit bravado and rationalise the exercise regime and disciplinary drill of moves to be followed. According to the Bulli Surf Club records, in 1921 the SLSA ruled that to compete, participants must hold a surf bronze medallion. In this way it was impossible to undermine the SLSA training programme. Through club competition and rivalry, surf lifesaving was turned into a type of 'sport' through events including belt races, beach sprints and surf races. As discussed by Pierre Bourdieu (2000), organised sport in the West is configured by Victorian bourgeois values, and is understood as an activity essential to the containment of masculine assertiveness, and integral to forging 'character'. Hence, through participating in SLSA club competitions as sport, the bodies of surf lifesavers had authority in these events to exhibit their masculine attributes and compare their physical strength as a feat of training regimes. Training, displaying and comparing through inter-club competition all provided opportunities to confirm eugenic discourses. In the Illawarra, during the first decades of the 20th century the SLSA appears to be valorised through the increases in the number of affiliated surf clubs, the growth in overall membership and participation in surf carnivals.

Under the territorialising impulses of the SLSA, the Illawarra beaches of the 1920s–1940s were discursively constituted by SLSA members as privileged spaces shaped for the performance of an Australian white heteromasculinity. The following two sections explore the narratives of surf club members. The first explores how surf club members reiterated hegemonic discourses of surf lifesaving. The second examines how surf club members also destabilised expected behaviours causing 'lifesaving trouble'. Together these sections explore the different ways surf club members negotiated the surf beach and the subject positioning of the surf lifesaver.

At the Surf Beach with Australian White Heteromasculinity

How Illawarra surf club members reiterated eugenic discourses of the SLSA was evident in 'Surf Notes' published in the *Illawarra Mercury* through intersecting themes: the athletic; family and; vigilance and service. The first theme of the athletic was evident in how most Illawarra club members adhered to the discourse of surf lifesaving as an exercise regime and emphatically performed their adherence to 'fitness'. Figure 5.1 illustrates how surf lifesaving bodies of the North Wollongong Surf Club, 1928, 'fit' a construction of masculinity within the eugenic discourses (Ramsland, 2000). To become a lifesaver, bodies needed to conform to exacting standards of the SLASA in terms of shape, fitness, firmness and overall appearance. The Illawarra Branch of the SLSA confirmed that the bronze medallion could be conceived as a 'body project' or 'makeover' for men:

> After passing, in most cases, a severe test, a course of probation must be served during which the recruit receives instruction leading to the examination for the Bronze Medallion – the proficiency award of the Association. On receiving this he becomes a qualified life-saver. (*Illawarra Mercury*, 1935g)

Surf lifesaving bodies were worked upon to 'match' SLSA ideals. Bodies were disciplined through the drills and training to produce flesh that was tanned and taut, adhering to the eugenic standards of 'healthy', 'attractive'

Figure 5.1 March Past of the North Wollongong Surf Life Saving Club (c. 1920) (*Source*: Unknown. From the collections of the Wollongong City Library and the Illawarra Historical Society)

and 'superior'. The Illawarra Branch of the SLSA constituted lifesaving as acceptable for men by presenting a particular version of masculinity that relied upon the dangers of rips and the physical force required to use lifesaving equipment including initially the belt, reel and line; then later, surf boats and surfboards.

The importance of the performativity of discipline and bravado to how bodies took their shape as lifesavers was illustrated in surf club reports circulated in the Illawarra. For example, writing under the pseudonym of 'Shark', the publicity officer of the Corrimal Surf Club wrote in October 1937:

> Weather again was unfavourable for surfing and stopped the beach being patronised by the surfers, but did not stop the members of our club. They possess the right spirit. (*Illawarra Mercury*, 1937k)

Similarly, 'Surf Ski' wrote in December 1937 that:

> Patrols last weekend were carried out under adverse weather conditions. However, nothing short of a cyclone will stop some of our members. (*Illawarra Mercury*, 1937l)

Surf lifesaving bodies were identifiable by their physical capacity and bravado in extreme weather conditions. Likewise, North Beach publicity officer, 'Sunshine' commented in the *Illawarra Mercury* in October 1931 that:

> Owing to the cold southerly on Sunday last the boys didn't participate in the surf ... It is sincerely hoped that members will take notice of the example shown by their club captain and five other members, who train every afternoon at the baths. (*Illawarra Mercury*, 1931e)

This report by 'Sunshine' hints at how the dynamics of pride and shame operated within the SLSA. Lifesaving was predicated on notions of gendered and aged shame. In this report club members were shamed as 'boys' who failed to secure distinction and legitimacy in the surf by not reproducing conventional gendered styles and movement because of a 'cold southerly'. Pride is mobilised through how the actions of the club capital reproduced SLSA gendered ideals regardless of the weather conditions. As argued by Elspeth Probyn (2000), the mobilisation of corporeal pride pushes the physical limits of what the body is capable of doing. In this case pride was mobilised through how the body of the captain reproduced a legitimate style of masculinity regardless of the weather.

The Illawarra SLSA archive demonstrated how members participated in competitive events, including surf carnivals. For example in the 1920s the 'Stevenson Cup' was organised by the Illawarra Branch of the SLSA for surf rescue and resuscitation (Middleton & Figtree, 1963). In the 1930s, the competitive surf carnival became popular. These events enabled lifesaving to reproduce on surf beaches a particular form of masculinity based on competitiveness, aggression and elements of conventional understandings of the Victorian sporting male. Competitive events organised by the SLSA became opportunities to display at the surf beach a particular version of white heteromasculinity as a 'natural' form of behaviour (Figure 5.2).

According to reports about surf carnivals in the *Illawarra Mercury*, the Illawarra beaches were important sites that promoted an aggressive, 'winner-takes-all' ethos. For example, publicity officer, 'Sunbaker', from Port Kembla Club reported in December 1935, that:

> Despite exceedingly cold weather last Sunday ... there were very few absentees amongst the competitors, although there were no spectators ... Well, 'stout fellows' of the surf clubs, we will see you all at Austinmer surf carnival on this Saturday week, to battle for honours in beach and

Figure 5.2 Winners of the Thirroul R & R Challenge Shield 1917, North Wollongong Surf Club (*Source*: Unknown. From the collections of the Wollongong City Library and the Illawarra Historical Society). Informed by ideas of masculinity informed by physical culture and sports, the wearing of trunks over the one-piece bathing costumes by the surf-lifesavers had the effect of enhancing the display of the size and shape of the male genitals

water events, and to see who's going to be the most prominent club of the season. (*Illawarra Mercury*, 1935h)

As evident in this report, competition enabled club members to participate in the discourse of war and sport, confirming accepted social understanding of the male identity. Similarly, in 1940, the Helensburgh/Stanwell Park Surf Club records described club members 'battling ... through pounding surf'. Lifesavers are described in terms of the masculine ideal of the 'tough guy' in the pursuit of heroism (honours). Surf carnivals, like other competitive sports discussed by Lynne Segal (1990), were understood as battlefields, organised spaces where men could channel their allegedly 'natural' aggressive instincts.

Furthermore, reports often conveyed the particular type of body required to win was one where the 'natural' aggressive instincts were disciplined by the SLSA. For example in 1935, the Illawarra Branch Annual Report heralded North Wollongong Surf Club as champions of the Illawarra Branch, winning the 'Illawarra Branch SLSA Championship Pennant', the 'Bligh Shield', the

'Klingtite Cup' and the 'Dawes Silver Reel'. The Branch record accounted for this achievement in the following statement:

> North Wollongong Surf Life Saving Club have asserted their superiority for the past 4 years in competition. To them we offer our congratulations for their successes which was due to the fine Club spirit, brought about by strict attention to drill, training and discipline. (Illawarra Branch Annual Report, 1934/1935)

This report clearly linked competition, winning and athleticism to adopting the SLSA rituals that schooled bodies in a military masculinity expressed through drills, physical training and discipline.

Winners of individual titles were given particular kudos in the 1930s. In Bourdieu's (1977) terms, surf lifesaving was a prime site for the acquisition of cultural capital. For example, Allan Fitzgerald (Corrimal Surf Club member and later North Wollongong Surf Club member) was constituted as an Illawarra hero in the surf club records following his win of the senior belt championship of Australia at Bondi in March 1936. The 1935/1936 Illawarra Branch Annual Report stated that:

> The only success that accrued to the Branch in these entries was the notable win by A. Fitzgerald, of the Senior Belt Championship of Australia ... we congratulate Mr. Fitzgerald on the honour he has brought to himself and to this Branch by carrying off this Championship.

Similarly, the Corrimal Surf Club records noted Allan Fitzgerald's achievement as 'the greatest honour this club and the Illawarra Branch has ever gained'. Likewise, the Corrimal Surf Club publicity officer condoned how training the body to conform to the normative bodily performance of the lifesaver was a way to acquire cultural capital and social status:

> On behalf of Allan Fitzgerald, let me thank all those who have so enthusiastically congratulated him on his fine performance in winning the belt championship of Australia. Corrimal Club is proud to number among its members such a champion, and we are proud to have so successfully represented the surf clubs of the Illawarra Branch. (*Illawarra Mercury*, 1936e)

On the one hand, the author mobilised pride to legitimise Corrimal Surf Club's claims over Allan Fitzgerald. At another level, pride was mobilised to legitimise the Illawarra Branch's positioning at the top of an Australian surfing hierarchy. The celebration of Allan Fitzgerald's achievement worked to reinforce and naturalise understandings of the surf beach as a heroic terrain, a test of 'manly virtues'. Participation in lifesaving at this time reinforced and naturalised the notion of masculinity that valued honour, training, discipline and tanning.

'Vigilance and Service'

The motto of the SLSA 'Vigilance and Service' was central to how the semi-naked bodies of lifesavers were gazed upon and dominantly read at the beach as neither sexual nor erotic. In the context of the surf beach, the surf lifesaver could parade his body clad only in a neck-to-knee costume without shame of being humiliated for deriving pride from bodily display because of how the SLSA had configured the beach as a masculine space. The surf understood as a threatening, dangerous space, helped to authorise the public's gaze over the ritualised and stylised codes of behaviour of the lifesaver. The drills, training and tests led to the illusion of the subjecthood of the lifesaver as a performance of vigilance and service. The surf beach became a privileged site in which semi-clad lifesaving bodies were performed, looked at, spoken about and gazed upon in terms of a voluntary public duty. The performance of the lifesaver, understood in terms of self-sacrifice for the safety of surf-bathers limited the potential for the public to gaze upon the visible physical signs of masculinity as sexual and instead sustained understandings of a respectable moral masculinity.

Honour, self-sacrifice and vigilance were central to what it meant to be a lifesaver. These characteristics were particularly evident in newspaper articles circulated during the 1930s. For example, in April 1935, an Illawarra Branch report highlighted in the *Illawarra Mercury* the charitable work of the SLSA:

> There is no material reward for the work. The voluntary spirit in which it was inaugurated has been preserved throughout. (*Illawarra Mercury*, 1935g)

The report then went on to present the statistics of SLSA rescues as a 'simple testimony of the efficiency of the clubs and the value of the movement to the community' (*Illawarra Mercury*, 1935g). Likewise, in a March 1937 edition of the *Illawarra Mercury*, a 'Surf Notes' article stated:

> A deep hole in the surf last week-end kept the members on patrol on the alert. However, the Surf Life Saving Association's motto of 'Vigilance and Service' was, as usual, well upheld and nothing serious eventuated. (*Illawarra Mercury*, 1937m)

Reference to the danger of 'a deep hole' naturalised the beach as masculine space. And, the desire to do duty promised public safety at the beach. Similarly, in October 1938, to legitimise surf lifesaving the Illawarra Branch mobilised pride at the number of lives saved, self-sacrifice and voluntarism:

> the essential fact that 338 lives were saved by members of surf life saving clubs affiliated with the branch is sufficient reason for us to be immensely proud of our achievements and certainly does justify our existence ... The appended list of rescues performed by the various clubs clearly demonstrates the value of surf life saving clubs, the ever growing popularity of surfing and the vigilance of the patrols. (*Illawarra Mercury*, 1938d)

Surf Notes published in the *Illawarra Mercury* emphasised lifesaving as an act of self-sacrifice through underscoring voluntary labour. Through this moral frame all acts could be used to convey a sense of righteous morality. For example, the secretary of North Beach Surf Club reported in December 1931 that even the voluntary construction of a drain was an integral part of the honourable labour of a surf lifesaver. He argued that club members 'are doing work voluntarily to make our beach one of the best' (*Illawarra Mercury*, 1931b). Similarly, in November 1939, the publicity officer for the SLSA Illawarra Branch reported that the newly formed Coniston Surf Club showed 'great enterprise' in their building endeavours which were all carried out by 'voluntary labour' (*Illawarra Mercury*, 1939b). In the same newspaper, Corrimal Surf Club reported that its club had 'the assurance of the club members that they will help to erect the new club room' at Corrimal (*Illawarra Mercury*, 1939c). Surf lifesavers were portrayed as model citizens through displays at the surf beach of honour, duty and service.

These results echo the findings of Booth (2001), Leone Huntsman (2001) and White (2007), who explored the positioning of lifesavers by members of the SLSA within the Sydney media. Semi-clad lifesaving bodies as an embodiment of a moral duty of self-sacrifice and vigilance were expected to display their physicality at the surf beach and had every right to make themselves visible in the 'public eye'. Indeed, as outlined by Booth (2001) following the mass rescue of 300 swimmers at Bondi, Sydney, in February 1938, the stature of lifesavers became enhanced to that of Australian national manhood. Given how the SLSA instilled specific notions of masculinity and citizenship it was perhaps not surprising that many surf club members joined the armed forces within the first year of World War II (Illawarra Branch Annual Report, 1943/1944). As discussed by Maxwell (1949) and James (1983) lifesavers were known to have even given displays during World War II. Australian club men held surf carnivals – Australian style – in Gaza, Tel Aviv and Jaffa, while bronze medallions were gained in Changi. The Changi POW SLSC was officially registered in SLSAA records.

Family

Booth (2001) and White (2007) have shown convincingly that the SLSA rules, training regimes and motto of 'Vigilance and Service' helped to fix and naturalise understandings of the surf beach as a heteromasculine space. But it was not just the performance of drills and ideas of self-sacrifice that reiterated heteromasculinity; the surf beach became bounded as masculine space through the layout and design of surf clubs, men and women's gendered division of labour, the celebration of the nuclear family and the remembrance of the intimate homosocial bonds of mateship. The identity of the lifesaver at the surf beach was celebrated as a member of both a normative heterosexual nuclear family household and an extended brotherhood forged by the social relationships of mateship. It is hard to overemphasise the historical weight that the ideology of heterosexuality had on defining the social relationships that comprised the surf club.

Archival records suggest that women took an active role in the Illawarra SLSA, despite being excluded from surf lifesaving because of how lifesaving was regarded as a masculine activity. During the early 1930s, women formed committees to organise and cater for surf club fundraising events. These events were integral to the financial and social relationships that sustained the surf clubs. Publicity officers regularly reported on the 'successes' of these events in the *Illawarra Mercury*. For example, in 1931 the publicity officer of North Beach Surf Club recorded that women 'conducted a dance at the Kiosk which was a huge success' (*Illawarra Mercury*, 1931b). Similarly, the publicity officer for Bulli Surf Club described the 'ladies' auxiliary' in 1938 as follows:

> The social spirit has been infused into Bulli Surf Club in a very definite manner, following the formation of a ladies' auxiliary ... The first of a series of fortnightly card evenings was conducted at the beach kiosk on Tuesday ... these card evenings will be the means of providing much desired pleasure for card players, and at the same time provide a good social spirit. (*Illawarra Mercury*, 1938e)

And 'Surf Ski' from Austinmer Surf Club reported that:

> The club dance held in the surf shed last Thursday, proved one of the best yet, and the social committee extends its thanks to all those who helped in making it so. (*Illawarra Mercury*, 1938f)

Within the SLSA the masculine/feminine binary was crucial to the construction of bodies and spaces. While men were involved in training, drills and rescue; the voluntary domestic work of women was essential to sustaining the social relationships and financial sustainability of Illawarra surf clubs. While women were conflated with domestic sociality, examining the role of women in surf clubs brings to the fore their importance in sustaining the surf lifesaving movement and extending understanding of the role of surf clubs in weekly, monthly and annual leisure activities.

Beach beauty pageants were also organised in the 1930s by Illawarra surf lifesaving clubs as both a leisure activity and important source of revenue. Australians were familiar with the concept of beauty pageants; competition to find Australia's most beautiful woman had been organised since 1908, initially by the *Lone Hand* magazine, then from 1926 by the owners of *Smith's Weekly* and *The Guardian Weekly*. The search for the perfect female form in the Illawarra was given feminine titles such as: 'queen competition' (*Illawarra Mercury*, 1930d); 'surf queen' (*Illawarra Mercury*, 1935i); 'popular girl' (*Illawarra Mercury*, 1938g); and 'surf girl' contests (*Illawarra Mercury*, 1939b). A condition of entry was that competitors were required to be 'local' to represent an Illawarra surf club (*Illawarra Mercury*, 1935i). Caroline Daley (2003) and Grant Rodwell (1999) argued that beach beauty pageantry was organised as celebrations of the 'eugenic family'. Local candidates were required to personify a 'healthy, racial ideal':

> In a few years, they would become perfect wives and mothers, their fit young bodies ready and willing to give birth. (Daley, 2003: 112)

However, it is impossible to ascertain from the archive if women participated because of the discursive framings of eugenics or the dominant forms of femininity associated with the intersection of discourses about fashion, youth, beauty and leisure. There is certainty that the names of all pageant entrants who aspired to a beautiful life were published in the *Illawarra Mercury*. However, major prizes were awarded to surf clubs rather than entrants. Hence, for example, in 1939 when Miss Jean Watson won the 'surf girl contest', Port Kembla Surf Club received a surf boat at £150 (*Illawarra Mercury*, 1939b; Middleton & Figtree, 1963). Parading bodies of women at the beach was clearly not portrayed in the media as unusual or indecent. On the contrary, the publishing of the names of participants in the *Illawarra Mercury* suggests that willingness to publically display bodies that conformed to conventional ideas of femininity was an achievement.

Many surf club records referred to the legacy of 'families'. For example, Corrimal Surf Club records stated that in 1932, Thomas Nubley joined the

Corrimal Club – 'the first of many Nubley's to go through this club'. Similarly, the Helensburgh/Stanwell Park Surf Club mentioned the Russel Family, with Geoff, Bruce and Jack, who were described as 'solid active members'. These records also outlined how Ted Shipton (Helensburgh/Stanwell Park Surf Club president from 1921–1930) was later joined by his sons 'to carry on the good work of Ted'. Another example from the Helensburgh/Stanwell Park Surf Club praised the Jardine family and in particular how the four Jardine brothers 'played an outstanding part in the various club activities, both in the administration and surfing side'. Further, the Helensburgh/Stanwell Park Surf Club outlined how 'fathers, sons and brothers' worked together voluntarily for the benefit of the surf club. Together these reports illustrated how surf clubs in the 1930s reinforced and naturalised notions of nuclear families and hegemonic heterosexuality. The Illawarra Surf Clubs were training grounds in respectable heteromasculinity in the 1930s.

The Surf Notes pages in the *Illawarra Mercury* in the 1930s also provided examples of how heterosexual identities were praised and normalised. For example, in November 1939, Wollongong Surf Club announced the marriage of Norm Shelton, the then club president:

> the heartiest of congratulations go to him and his bride, who was Miss Connie Quilty. (*Illawarra Mercury*, 1939d)

In the same newspaper Wollongong Surf Club also congratulated Mr and Mrs Bill Armitage 'on the advent of another surfer to the family' (*Illawarra Mercury*, 1939d), while North Wollongong Surf Club articulated: 'Congratulations to Fitz and Mrs Fitzgerald on the arrival of a bonny son' (*Illawarra Mercury*, 1939e). Lifesavers were praised publically for engaging in 'hegemonic heterosexuality'. A respectability to surf lifesaving and individual members, accrued through announcing those who conformed to hegemonic heterosexuality through marriage and becoming fathers. Furthermore, these announced further naturalised gendered assumptions about how only sons would become future 'surfers'.

Announcements of births, deaths and marriages of surf club members in the newspaper did not only illustrate how the SLSA made sense of the gendered and sexed family lives of lifesavers but also the homosocial bonds of mateship. For example in November 1939, the *Illawarra Mercury* reported the remembrance ceremony for Bill Curtiss, a former Bulli Surf Club member. The bonds of mateship were central to the service:

> A most impressive ceremony took place in the Bulli Cemetery last Sunday, when the members of the Woonona Surf Life Saving Club kept alive the memories of their late club mate, Bill Curtiss, who passed away

> 12 months ago. Members marched from the entrance to the grave, where a beautiful glass covered wreath . . . done in the club colours of white and black was placed on the grave by the president, Mr. T. Medlyn. (*Illawarra Mercury*, 1939f)

Remembrance ceremonies were an integral part of the performative of a respected surfing masculinity and culture of mateship. The social bonds of mateship reinforced and naturalised the culture of mateship as an expression of care for 'mates'. Remembrance services mobilised pride from performing a respectable lifesaving masculinity that relied upon prioritising a duty of care over physical strength.

The important points to emerge from the Illawarra surf club records were insights into the process by which scantily clad bodies of lifesavers belonged on the surf beach. Like in Sydney, the naturalised presence of men-as-lifesavers relied upon understanding the beach as a masculine space and bodies of lifesavers for their moral rather than physical capacities. The bodies of lifesavers were disciplined by club captains and the SLSA tests, training regimes and drills. However, it would be too simplistic to read the lifesaving bodies as compliant with normative behaviours that governed normative constructions of masculinity for lifesavers. The next section explores the surf club records for evidence of how the hegemonies governing normative constructions of lifesavers' masculinity were both affirmed and contested through tanning and surfing.

Lifesaving Trouble

How lifesaving bodies took on public identities fashioned by moral masculinities was crucial in understanding how the surf beach was framed as ostensibly non-sexual. An important element of this was the disciplining and surveillance of the surf beach and of lifesaving bodies through the representation of lifesavers and lifesaving clubs in militaristic terms: battles, honours, heroes, discipline, training and service. Lifesavers that refused to follow the SLSA script of masculinity caused 'lifesaving trouble'.

For example, consider the trouble caused by tanning. On the one hand, tanned bodies were integral to legitimising surf lifesaving, through positioning lifesavers as fearless Greek gods (White, 2007). On the other hand, the pleasures of tanning were also understood as central to a hedonistic culture. In February 1931, 'Beltman' wrote a Surf Note to the *Illawarra Mercury* that outlined how some club members prioritised tanning over training for competitions:

> It only goes to show what can be done if the spirit is in the game, and some of our members would do much more good for the Club, if they put

> a bit more energy into the surf carnival events instead of lying on the beach all day. (*Illawarra Mercury*, 1931f)

'Beltman' understood these tanning bodies as lazy. Lying on the beach all day was understood by Beltman as taboo, breaking what was deemed proper behaviour for lifesavers.

However, the following bathing season a Surf Note from a surf club publicity officer writing under the pseudonym 'Sunshine' expressed his support for tanning:

> Owing to the cold southerly on Sunday last the boys didn't participate in the surf, but took to the next best thing – 'Sun Baking' in the club room. (*Illawarra Mercury*, 1931e)

For, 'Sunshine', 'the next best thing' to training in the surf was the pleasure of tanning. Other publicity officers sanctioned the hedonism of sun-baking. For instance, in September 1935, writing under the pseudonym 'Farmer', the North Beach Surf Club publicity officer announced that: 'It will not be long before we are enjoying the surf and sunshine' (*Illawarra Mercury*, 1935j). Indeed, the pseudonyms 'Sunbaker' and 'Sunshine' suggest that tanning was sanctioned in some surf clubs.

Equally, lifesaving trouble was created by surfboards. The SLSA legitimised the use of surfboards as a rescue tool from the 1930s. However, lifesavers clearly enjoyed surfing and demonstrating their surfing skills in public. For example in January 1936, one Port Kembla Surf Club member, named Fred, was reported in the Surf Notes of the *Illawarra Mercury* as always being with his surfboard:

> The boys say that Fred and his surf board are inseparable, and that if it wasn't so heavy he would take it home and sleep on it. (*Illawarra Mercury*, 1936f)

In October 1938 the publicity officer of Austinmer Surf Club reported that: 'Quite a number of the boys have been seen brushing up their acts on the surf boards this week' (*Illawarra Mercury*, 1938h). Perhaps the most notorious was the Corrimal Surf Club lifesaver, Walter Dare, who won two trophies for his club in 1937, yet also created a spectacle of himself by riding a surfboard in a 'head-stand' position in the late 1930s (Corrimal Surf Club records). Again, the pseudonyms of the publicity officers, 'Splash' and 'Surfo', suggest a light-heartedness and point to the pleasures of playing in the surf rather than to be always disciplined.

The Surf Notes published in the *Illawarra Mercury* can be read as attempts by surf club publicity officers to police boundaries for surf lifesavers so their scantily clad bodies did not cause affront to the public or authorities. Yet, publicity officer's narratives clearly suggested that for some club members the surf beach was as much a homosocial space of hedonistic pleasure as a training ground in an acceptable version of heteromasculinity that resonated with the sacredness of nationhood. These findings complement the work of Jaggard (1997) who also explored the trouble caused by lifesavers' reported lived experiences when compared with the myths of the SLSA. He positioned lifesavers as 'chameleons in the surf' to highlight how surf club members were constantly negotiating understandings of improper and proper masculine behaviour at the surf beach. Jaggard argued that 'discipline and larrikinism were, and still are, opposite sides of the same surf lifesaving coin' (Jaggard, 1997: 190). Over the next decades, men who sought to establish an acceptable male identity as defined by their peer group through surfing distanced themselves from the conformist ideals of the SLSA. These men did not wish to play the role of the SLSA lifesaver or 'clubbies'. Instead, these men sought to express their own identities in and through the surf. This was made possible from the 1940s, with aesthetic surfing displays enabled by the lighter 'Malibu' board. Kent Pearson (1979) documented the competing territorial claims between the two surfing cultures; the 'clubbies' and 'surfies'.

Conclusion

As this chapter shows, the case of the surf beach was illustrative of Foucault's notion of disciplinary power, and how Judith Butler understood gender as the repeated stylisation of the body to produce the appearance of a subject. The chapter began with the assumption that there is nothing inherently masculine about the surf beach. Attention was given to how the Australian surf beach became taken for granted as a masculine space from the 1920s through the regimes of truth, power and subjectivity of the SLSA. Working within the heteronormative alignment of sex and gender, the SLSA fashioned the lifesaver within a particular gendered construction of masculinity that effectively excluded women from becoming full members until the 1980s. Within SLSA-affiliated surf clubs women almost exclusively performed a domestic femininity through organising fund-raising activities or emphasising femininity of the surf carnival beauty pageant. Acceptable displays of femininity at the beach became defined in relationship to the heteromasculinity of the surf lifesaver as 'master of the beach'.

The Illawarra Branch of the SLSA and the publicity officers from each SLSA affiliated surf club helped to establish particular expectations about how the public should gaze upon the scantily dressed bodies of surf lifesavers. As Butler (1997: 5) argued, it is 'within the terms of language that a certain social existence of the body first becomes possible'. Lifesavers were framed through language as the embodiment of a moral masculinity rather than as sexual or feminine. This framing required the disciplinary power of the intersecting discourses of the athlete, the nuclear family and self-effacement.

The discourse analysis also established how surf lifesavers constantly negotiated their gendered identities in and through the spaces of the surf beach configured by the SLSA. Surf club members were instrumental in stabilising the boundaries of the surf beach as a masculine space through managing their corporeality both as individual and social bodies along the lines authorised by the SLSA. At the same time, the regulatory practices of the surf beach were integral to how surf lifesavers were continually being performed. Surf club records confirmed understandings of surf lifesavers as being accorded with the greatest moral, physical and social stature of all surf-bathers. And yet, this regulatory fiction does not remain secure. In the surf records there is substantial evidence of the hedonistic pleasures of surf lifesaving. Lying around on the beach tanning disrupted the proper public identities of lifesaving masculinity. Similarly, surfboarding for pleasure offered a different way of doing surfing, and created lifesaver trouble. This trouble is central to understanding Australian surfing and the territorialisation of the surf beach by 'surfies' from the late 1940s onward (see Booth, 2001; Jaggard, 1997).

6 Making Bathing 'Modern'

So much attention is paid to appearances nowadays that surf brassieres are worn by most women. They are of net, and hold their figure firmly beneath the suit. The slim line of the figure may be preserved by a sport's girdle, fitting closely round the abdomen ... When the surf suit is donned over this, the effect is very good, and there is the satisfaction of knowing one's figure is well controlled

Illawarra Mercury (1929j)

Surf-brassieres and sports girdles are described in this article appearing in the *Illawarra Mercury* in December 1929 as essential for fashionable bathing bodies to appear 'well controlled' on the surf beach. At one level the article illustrates how the beach became framed within a culture of consumption as leisure time increased and the availability of consumer goods grew. The growing economies of cities based on industrial capitalism facilitated the emergence, standardization and synchronisation of leisure time and commercialisation of beach culture. Here families and religious authorities were less able to oversee the behaviour of young people. Instead, the state and corporate capitalism extended their material and cultural control over daily life, including the co-ordination of different forms of mobility that together helped to stabilise the boundaries of regions and nations.

At another level the individual beach goers who wore the surf-brassier and sports girdle beneath their bathing suit complied with hegemonic gendered constructions of beauty and fashion in the apparel industry. In the late 1920s the stylisation of bodies through wearing surf-brassieres and sports girdles enabled the 'slim line' bathing bodies to be made visible as fashionable, feminine and sexual. At the beach, women were increasingly portrayed in advertisements as sexual beings. On the one hand, the women depicted broke out of the image of sheltered domesticity and purity, on the other hand newspaper articles and advertisements for bathing apparel increasingly depicted women as valuable not for their childbearing and domesticity, but principally as sexual objects. Looking at bathing bodies, however, was no longer framed as an illicit pleasure by the market. Instead, bodies disrobing at the beach became a principal source of social status of becoming modern women, knowing they were going to be carefully scrutinised in terms of their size, shape and form. Industrial society brought a profusion of goods and advertising in consumer culture. Beach bodies became symbolic of a 'valued' life. Styles of consumption became one way for individuals to perform their identities as young, cosmopolitan, fashionable and modern as well as obtain social distinction.

This chapter is divided into five sections: bathing, swimming, fashionable, swimsuit and tanning bodies. Each section casts light on how in the early decades of the 20th century the conventions of genteel Victorian culture become constituted by some bathers and swimmers as out-dated. The Illawarra archive material is supplemented in this chapter by articles from *The Australian Women's Weekly (first published June 10, 1933)*. Advocates for change drew on rational discourses of science and physical culture and romantic discourses of heterosexual love to distance themselves from the Victorian past and what they considered dysfunctional sexual repressiveness as exemplified in the moral policing of the bathing reserve (see Chapter 2). Instead, they presented themselves as pioneers of a 'modern' beach culture. Bathing ordinances dating from 1906 in New South Wales aimed to restrict sexual culture in public by regulating bathing costumes. Those who advocated for a 'modern' beach culture reorganised heterosexual culture by drawing on different framings of the beach. Those within swimming institutions often used scientific framings and made bathing modern through eugenics and heliotherapy. For those advocating scientific framings of the beach, swimming bodies became positioned as non-sexual and as 'good' citizens. Others framed the beach through consumer culture, and made the beach modern by celebrating the beach through international fashions and celebrities. Framed by consumerism the beach was portrayed as a legitimate space of public intimacy and coupling. The beach was represented as a publicly accessible glamorous heterosexual culture, a (hetero)sexy space. Men and women were able to assert a heterosexual identity and agency. Conservatives, who were labelled as 'wowsers', 'Mrs Grundys' or 'Spoonerites' (after Mr Eric Spooner's amendments in 1935 to Bathing Ordinance No. 52 of the Local Government Act 1919) thought of the public sexual culture of the host spaces of the beach as indecent. In the first decades of the 20th century the beach was a site of different expressions of heterosexuality. Philip Hubbard noted that: '[t]here appears to be little overt consideration of how moral heterosexual performances are naturalised in a variety of "everyday" social settings, either "public" or "private"' (Hubbard, 2000: 206). This chapter examines how bodies were a site of negotiation and struggle over what were regarded as moral and immoral expressions of heterosexuality at the Australian beach in the first decades of the 20th century.

An overarching theme of the chapter is that while there was different framings of what constituted moral and immoral heterosexualities in the first decades of the 20th century, much remained the same at the surf beach with regard to how bodies of men and women were gendered and sexed. The chapter explores that while women were making gains in the world beyond the family, the heterosexual-oriented strains of Australian beach culture were functioning to claim more control over women's bodies.

Bathing Bodies: 'Sexes Mingling Together as Nature Intended'

As discussed in Chapter 2, drawing on bathing ordinances, local authorities in New South Wales passed by-laws to regulate beach use, including prohibiting men and women from bathing together. As argued in Chapter 2, there was a sexual imperative to bathing ordinances. Genitals of the opposite sex were not to be visible. Men were made to feel unmanly if caught looking at naked female flesh. Shame about the self and body also discouraged 'respectable' men from posing naked, covering their genitalia to protect the sensibilities of women. Equally, bourgeois women were well-schooled in the humiliation that arose from displaying their flesh in public. However, newspaper reports in the Illawarra suggest that many bathers defied the rules that denigrated the naked flesh and segregated the sexes. Despite non-sexual imperatives of bathing ordinances to regulate the sexual gaze at bathing reserves, newspaper reports in the *Illawarra Mercury* suggest that since 1869 the beaches of Wollongong had a reputation for 'mixed-bathing' and the implicit eroticism encoded in such a site. For example, as reported in the *Illawarra Mercury*: 'The establishment of bathing places for both sexes in Wollongong has been a question frequently brought before the public, but hitherto nothing has been done' (*Illawarra Mercury*, 1869).

Those who advocated for two-gender exclusiveness of the bathing reserve argued that sexual motives and erotic interests of adult men drove the pleasures of mixed-sex bathing in Wollongong. Mixed-bathing was about prohibited pleasures (Figure 6.1). For example, in 1907 an anonymous writer in the *Sydney Morning Herald*, discussing surf-bathing in the Illawarra stated that: 'mixed bathing lowers the morals of the people and has a tendency to animalise the race' (*Sydney Morning Herald*, 1907, cited in Huntsman, 2001: 59). This author points to segregation along the lines of gender and race as a necessity to maintain bathing as an exclusively non-sexual form of pleasurable activity. The author goes on to speculate that men and women bathing together in the Illawarra involved illicit sexual encounters that undermined civility. Mixed-bathing broke the boundaries of acceptable behaviour, blurring the cultural distinction between 'humans' and 'animals'. Mixed-bathing

Figure 6.1 Bathers at Thirroul Beach (c. 1915) (*Source*: Unknown. From the collections of the Wollongong City Library and the Illawarra Historical Society)

was equated with animalistic sexuality. Like the dance halls and movie theatres, mixed-bathing was one example of changing courtship patterns that were believed to undermine a respectable femininity by occurring outside the parental supervision of the home.

Similarly, an anonymous letter to the *Illawarra Mercury* editor in 1908 referred to a 'mixed bathing club' in Wollongong with special disdain, shaming men.

> I understand they are canvassing the young girls of the town for the purpose of joining their mixed bathing club. I am sorry they are not sufficiently endowed with more manhood that they might know that it is base impertinence on their part to approach respectable young girls in such a matter ... There is a growing looseness amongst our young people of both sexes that is deplorable ... Will we be able to maintain our country with the shattered remnants when the time comes to repel an invasion which may not be far distant. It is to be hoped that our young girls who have self respect will not yield to the bland young 'gentlemen' who are interesting themselves in a cause of degeneracy for them. The oft-quoted axiom 'Righteousness exalteth a nation,' the city, or individual. This we could profitably ponder over. (*Illawarra Mercury*, 1908b)

Drawing upon biblical authority, this writer was offended by young men who encouraged mixed-bathing, positioning them as both impertinent and degenerate. Mixed-bathing was understood as having some sexual component, referring to 'a growing looseness amongst young people', and therefore violated the inherited Victorian bourgeois moral codes of respectable masculinity and femininity. Men who bathed with women in the surf became embodiments of shame. Conceived as being deficient from male norms, the author questions their masculinity. Men who bathed with women in the surf became targets of shame because they were 'not sufficiently endowed with more manhood' and were 'bland young gentlemen'. The author also reminded readers that decent, respectable young women do not bathe publically with men in a semi-naked state. Women who bathed in public were also to feel shame. This author mobilised shame to socially devalue surf-bathing bodies that did not conform to bathing segregated along the lines of sex. This author went on to suggest the demise of this control by surf-bathers of their sexual bodies would lead to the demise of the social body and the downfall of the new Australian nation. This anonymous writer aligned their fear of a lack of control over sexuality with their anxiety that the younger generation lacked the discipline and sense of civic duty to defend the new Australian nation. In conjunction with the Federation of the nation and emergent nationalism, nothing less than protection of the new Australian nation was apparently at stake in the pleasures of mixed surf-bathing.

In contrast, men who had enjoyed the delights of mixed-bathing positioned such complaints as those belonging to 'wowsers', who sought to ban everything pleasurable. Letters to the editors from men who advocated mixed-bathing emphasised the frames of purity and naturalness. In the context of discussions regarding building a new ocean bath in Wollongong for women an anonymous male writer suggested increasing the size of the 'gent's bath to accommodate both sexes'. He based his justification following a visit to Katoomba, the Blue Mountains, New South Wales, where he had experienced the pleasures of a communal spa bath. At this time, most people could not swim, therefore, bathing at spa resorts was positioned primarily about health:

> the bath for mixed bathers was full of ladies and gents enjoying life as it should be enjoyed. The air was full of the ringing laughter of both sexes mingling together as nature intended it to be. What a contrast to Wollongong, where the ladies are in quarantine on Flagstaff Hill! (*Illawarra Mercury*, 1917)

Rather than cultural codes of the obscene or vulnerability and personal privacy, communal bathing between genders was positioned as 'enjoying life . . . as nature intended'. The author stressed that men and women bathing together were displaying their natural state. To maintain respectability, the pleasures of mixed-bathing was positioned as natural, to counter arguments of mixed-bathing as impure. Pleasure was derived in purportedly non-sexual ways. The author draws on wider arguments that mixed-bathing prevented rather than encouraged illicit sexual behaviour. Following this argument, mixed-bathing helped to undermine eroticism and the hypocrisy, shame and secrecy surrounding sex. For this author, women who bathed in isolation were positioned as unnatural. This author positioned mixed-bathing as a healthy, modern and natural leisure practice.

In the 1920s Wollongong Council did make provision for men and women to bathe together with the construction of the Continental Baths. In Wollongong, the demand for a bath that allowed men and women to bathe together was positioned as having European rather than British origins. However, it was not the arguments that mixed-bathing helped to undermine the eroticism surrounding the bathing body that convinced aldermen. Instead, it was the physical culture of swimming as an institutionalised sport. Public pressure from newly formed swimming clubs was used to justify the expenditure in terms of ratepayer's health, convincing Wollongong Council to build a new Wollongong Central Baths that offered mixed-bathing sessions.

Swimming Bodies

From the 1920s, Wollongong Council followed in the wake of other Australian cities and country towns in the provision or maintenance of swimming pools (Lewi & Nichols, 2011). During the 1920s, volunteers played a central role in the construction of swimming pools in the rock-platforms at Coal Cliff, Scarborough, Wombarra, Austinmer and Bulli (Hutton, 1997). These swimming pools were built primarily by coal miners for their families. As discussed by Douglas Booth (2001) and Caroline Daley (2003) swimming at the turn of the 20th century had become mainstreamed in Australia, New Zealand and Britain through eugenicists, competitions and celebrities. In Australia, swimming clubs and competitive displays of swimming – the swimming carnival – had existed since the 1880s, however, few people were able to swim. In 1909 the Amateur Swimming Union of Australia was founded in an exclusive and respectable Sports Club on Hunters Street in Sydney. Drawing on claims of medical doctors and eugenicists, swimming was framed in terms of executing a rational, sensible form of exercise that aided circulation, purified the body and strengthened respiration. Swimming bodies were positioned in terms of 'good' Australian citizenship for taking responsibility for their own physical health and having the capacity to save the lives of others in danger of drowning. For some, swimming bodies were made respectable and non-sexual by how the sport of swimming reshaped bodies with discourses of physical culture. Swimming bodies were understood as both beautiful and healthy.

Swimming also began to be appreciated in terms of producing celebrities and national heroes, like Annette Kellerman, Barney Kieran and Frank Beaurepaire. Annette Kellerman, swimmer and vaudeville entertainer, led the transformation of social norms surrounding semi-clad swimming bodies in the early 1900s. In 1902, Annette Kellerman established a New South Wales record time for the 100 yards (91.4 metres) and a world record for the mile (1.6 km). Annette Kellerman used her swimming talents to entertain crowds in Australia and New Zealand and later in Britain and the United States (Woollacott, 2008). She toured Australia as part of a vaudeville troupe, Mr Fred Kellerman's Comedy Company, as well as giving exhibition swims to entertain audiences at swimming carnivals. One exhibition swim was held in the gentlemen's baths in Wollongong in February 1902. The *Illawarra Mercury* reported in January 1902 that:

> Miss Annette Kellerman, who appears with Mr Fred Kellerman's company on the 18th inst., is a very distinguished young swimmer. Besides winning the ladies championship handicap race at the East Sydney Carnival (from scratch), did the high dive for ladies (50 ft. high) and has since lowered the English ladies record (39 mins., 52 secs.) for one mile, this lady covering the distance in 34 mins., 20 secs. It would be rather interesting to our local swimmers to see this young mermaid perform. (*Illawarra Mercury*, 1902a)

A later edition of the *Illawarra Mercury* gave more details of the exhibition swim:

> It was announced that the company [Kellerman Comedy Co.] would again appear at Wollongong on February 15th, the date of the swimming carnival when Miss Kellerman will give an exhibition at the gentlemen's baths. (*Illawarra Mercury*, 1902b)

Annette Kellerman competed in swimming carnivals wearing men's racing swimsuits that revealed half her thighs. Audiences therefore saw more naked flesh than generally allowed in public. By the standards of the day, her swimsuit was risqué. Indeed, in 1907, international publicity was generated by her arrest on a Boston beach in the United States of America for indecency. She had worn one of her swimsuits that was skirtless and that clung to her body and exposed her thighs. Annette Kellerman drew on her athletic abilities to justify why she was required to reveal more flesh than other women. Audiences and newspaper reporters at swimming carnivals, like the judge in Boston, accepted her defence that the cumbersome women's bathing costumes worked against the health benefits of swimming. As an exponent of physical culture she became positioned as a role model of respectable womanhood. While Annette Kellerman displayed her body for a living, her body was clothed in the respectability of an athlete that gave audiences permission to gaze closely without sexual shame. The institutionalised context of swimming as a sport attempted to work against any pleasure or eroticism associated with the emphasis on the female body as spectacle. This is illustrated in the report published in the *South Coast Times* in February 1902 of Annette Kellerman's appearance at the swimming carnival in Wollongong:

> On Saturday afternoon a very large crowd of spectators assembled in the vicinity of the men's baths, Wollongong, to witness the carnival that had been arranged. The principal attraction was the exhibition of swimming, diving, etc., by Miss Kellerman. This young lady won hearty plaudits all around for her many clever natatorial [swimming underwater] feats. (*South Coast Times*, 1902)

Permission to gaze at semi-naked bodies at a swimming carnival was respectable when understood as a professional display of swimming talents; while participation in the physical cultures of swimming was increasingly understood as a mark of Australian citizenship.

Annette Kellerman turned her faith in swimming and her swimming toned athletic body into a marketable commodity. She promoted the physical benefits of swimming in her books *How to Swim* (1918) and *Physical Beauty and How to Keep It* (1919). In these books she positioned swimming as her saviour from a crippling childhood illness. Swimming lay at the foundation of her strength, health and beauty. Swimming as an exercise programme became a project that reshaped the body according to ideas of beauty at the time. In the 1920s, she launched her own line of one-piece bathing suits, the 'Annette Kellerman's'. The commercially available swimsuit had a tight-fitting skirt (or modesty panel) which came to just above the knees over the existing swimsuit. Semi-clad swimming bodies were no longer understood as indecent when framed in terms of the athletic, the good of the nation or improved health.

In Wollongong, work began in 1924 on the Central Baths or Continental Baths (because, like at French bathing resorts bathing would not be segregated

Figure 6.2 Continental Baths, North Wollongong Beach (c. 1930) (*Source*: Unknown. From the collections of the Wollongong City Library and the Illawarra Historical Society)

by sex) (Figure 6.2). In keeping with Wollongong Council's outlaw of facilitating leisure and tourism cultures as discussed in Chapter 4, the baths received only partial funding. The *Illawarra Mercury* printed the names of volunteers who worked on the new construction. The open-air, salt-water pool was built on the foreshore. Swimming bodies could enjoy the health benefits of the sea air and saltwater, but were shielded from dangerous riptides. In March 1928 the baths were opened. The Wollongong Amateur Swimming Club organised a 50-yard ladies' race as part of the opening swimming carnival. While costume stewards at swimming carnivals enforced the rules that 'lady competitors' made their way to the starting blocks, and left the pool, covered by a gown, swimming events provided a circumstance for semi-naked bodies to be revealed in public. An estimated 4000 people attended the opening swimming contest, signalling the importance of swimming carnivals as a form of entertainment. Reports in the newspapers did not speculate if the popularity of the opening ceremony was because spectators took pleasure from viewing swimming bodies in apparently non-sexual ways, as role models of health and beauty. Instead, a reporter for the *South Coast Times* considered the baths 'not only a credit to the town, but also of great public benefit' (*South Coast Times*, 1928). While, the Mayor (Alderman N.M. Smith) was reported in the *Illawarra Mercury* as having said: 'they [baths and sheds] were one of the most progressive moves, from the Council's point of view, in the town' (*Illawarra Mercury*, 1928). In this case, progress referred to allowing men and women to bathe together. A year later an article in the *Illawarra Mercury* reported on the popularity of the Continental Baths:

> All day long a steady stream of people visited the baths, but at eight o'clock the evening crowd was at its maximum. At that hour the surface of the baths was a mass of bobbing heads and splashing legs and arms, while to walk around the sides and on the space in front of the dressing sheds was almost an impossibility. (*Illawarra Mercury*, 1929k)

During the 1920s the Royal Life Saving Society held swimming and life-saving lessons in the Continental Baths for school girls and boys. The Continental Baths also staged the annual swimming carnivals of Christian Brothers' College, The Wollongong High School, Junior Technical School and the Wollongong Amateur Swimming Club. At the Continental Baths bodies could share the enclosure because men and women were brought together primarily by swimming and a new regime of power, truth and subjectivities. At the Continental Baths men and women were 'disciplined' to fit within prescribed gender norms of swimming. Within the context of the social relationships of swimming, the scantily clad body could be legitimately gazed upon in ways which were ostensibly non-sexual, including discourses of health, training and nationalism.

Nevertheless, during the 1930s some women still endorsed the sex segregation of bathing bodies:

> There are yet quite a lot of people who refrain from bathing in the Continental bathing area, and for those people surely the Council can give a little attention. The bathing shed at the ladies' baths requires attention, and I feel that the Council might spare a little time and attention for those respectable people who have not become modern enough to display themselves to the view of the general public. (*Illawarra Mercury*, 1932)

This anonymous bather acknowledged how the bathing body at the Continental Baths had become a source of pleasure, as something to be displayed to approving others. She positioned these pleasures as 'modern' attitudes towards the bathing body and sexuality. The author stigmatises the appearance of these bathing bodies in public as unacceptable. Some women felt unease at exhibiting their bodies in public at the Continental Baths. Instead, they wanted to retain a segregated bathing space where they did not come under the heterosexual gaze of men.

Other women wrote to *The Australian Women's Weekly* complaining how their movement at the beach was restricted and how they were constantly being positioned by heterosexual men – who they termed the 'pirate pest' – as sexual. For example, Mrs T. Mitchell wrote:

> With the approach of summer comes again the 'pirate' pest. I do think it time something was done. Why doesn't the law make this a criminal offence? It is almost impossible to walk alone on the beach front where I live [Bondi] without being accosted by a prowling pedestrian, and very often embarrassed. Is it fair that women, because they are unescorted, should suffer these indignities? (*The Australian Women's Weekly*, 1934a: 21)

As argued by Gill Valentine (1989: 389) 'women's inhibited use and occupation of public space is therefore a spatial expression of patriarchy'. Australian beach culture was not only raced and gendered but also sexed. As Mrs T. Mitchell's letter illustrated, heterosexual women could not claim entitlement to the beach in the same way as heterosexual men.

M. White responded to Mrs Mitchell's letter in *The Australian Women's Weekly* by qualifying her argument. She reprimanded women who made themselves visible at the beach clad scantily in bathing costumes, while praising swimming bodies:

> Mrs Mitchell (15/12/'34) is rather too sweeping in her charges against men in their alleged 'piratical' methods on the beaches. The average girl, who haunts the beaches, either singly or in packs, is all out to capture the attention of the male. It does not take an ultra scanty bathing costume to declare the type of woman within – personality shines from even the stuffiest of gowns. Many girls having nothing in their heads but stupidities – they court the attentions of the males of a corresponding type and then bleat sob stuff about being insulted. Even the craziest of young men know their marks well enough and don't attempt to molest young women who are obviously on the beach because they enjoy the surf, the fresh air, and the sunshine. (*The Australian Women's Weekly*, 1935a: 19)

M. White differentiated between respectable and disreputable femininities. M. White promoted the virtues of self-discipline of a beach culture configured by the discourses of health. For M. White, any claims on the beach by women should be mediated through the desire for respectability and safety. M. White judged the actions of women who did not police themselves accordingly as irresponsible, hypersexual and putting themselves at risk by wearing 'ultra scanty bathing costume[s]'. At the beach women had to continually negotiate how to be recognised as worthy respectable citizens.

Fashionable Bodies

This section explores further how swimming, bathing and beach bodies on display at the beach were caught in cultural practices that objectified and sexualised young women. Consumption of transnational beach fashions was fundamental to enabling young women to enter the beach. At the same time, transnational beach fashions generated a particular form of female visibility, restricting expressions of difference within and between women. Young women were addressed as consumers, with desires to be fashionable, glamorous and hailed at the beach by the male heterosexual gaze as sexually attractive. It became natural for young women to display their bodies at the beach in the latest swimwear. As Laura Fraser (1994) argued, however, the wearing of a one-piece or two-piece bathing costume at the beach necessitated an 'inner corset' that was no less repressive and restrictive than donning a corset or girdle.

Victorian social etiquette had always encouraged women to bathe at resorts fully clothed. In the 1890s, bathing resort fashion imported from England normally required women to wear a three-piece suit at bathing places: a serge jacket, loose drawers that tied below the knee, and a weighted skirt worn over the drawers. When the New South Wales neck-to-knee bathing costume ordinance was introduced in 1906, many women were already wearing far more clothing to bathe than demanded by the legislation, including shoes, stockings and corsets. Bathing fully clothed in private was a necessity to prevent a young women's moral downfall by arousing the sexual agency of men. An anonymous letter to the editor of the *Sydney Morning Herald* by 'Daily Dipper' in February 1907 reflected this sentiment:

> I am an Australian girl, sharing our national love for the water. I am no prude, and think the human figure the highest form of beauty, but no true woman would exhibit herself to all who care to gaze, clad in the thinnest and tightest of gowns. (*Sydney Morning Herald*, 1907d: 5)

Scantily clad young women supposedly presented a danger to society by encouraging masculine virility. As argued by Daley (2005) at the start of the 20th century semi-naked bodies of women belonged in the chorus line of stage productions or on pornographic postcards. Three decades later, displaying swimming bodies was understood as a common, even desirable tourism and leisure activity.

At the beach resort, rather than acting shy and demure, young women were increasingly expected to act in a manner that was glamorous, fashionable and sexually desirable. The increased popularity of the practice of movie theatre going was also integral to the performativity of modern femininity at the beach. In the late 1920s, as Daley (2003) pointed out, Hollywood had become for many young people the measure of glamour and beauty. Regional and metropolitan Australian audiences gazed upon pictures of glamorous movie stars in movie theatres, newspapers and magazines. In turn, their celebrity status was used to sell beach resort fashions. For example, in December 1929, the front page of the *Illawarra Mercury* (1929j) featured the Hollywood actress, Nancy Carroll, reminding readers she was a 'Paramount Star' in the film *Sweetie*. She was described as wearing a 'Chic Surf Ensemble'.

Hollywood stars were deployed to convince young consumers of the cosmopolitan qualities of beach wear both in and beyond the metropolitan centres of Australia. Such marketing efforts enabled adaptations of glamorous Hollywood images to a regional Australian audience. Dressing up for the beach in the transnational clothing fashion industry provided possibilities to play out the gendered roles of glamorous American movie stars at Australian beach resorts. Drawing on Judith Butler's (1990) theoretical constructs of performativity, the regulatory regimes of Hollywood glamour prohibited and enabled bathing bodies to perform in specific ways at the beach. In the 1920s Hollywood movie stars helped challenge the bourgeois traditions, habits, routines and dress codes of the beach. A new group of enthusiastic young consumers were willing to experiment with totally new ways of dressing up for the beach inspired by beach resort fashion from North America, Britain and Europe rather than swimming. Unlike Australia, municipal authorities in these countries no longer enforced neck-to-knee bathing costume regulations.

Fashion reporters for the *Illawarra Mercury* and *The Australian Women's Weekly* meant that beach wear transcended national boundaries. Ideas about beauty, romance, love and fashion flowed in from overseas reporters. The fashion industry courted young heterosexual consumers, reconfiguring notions of femininity but challenging conventions of respectability with 'daring' designs. Likewise, *Illawarra Mercury* advertisements stated that the swim wear designs were considered 'youthful' (*Illawarra Mercury*, 1929l) and 'daring' (*Illawarra Mercury*, 1929m). For example, in 1933, Nell Murray – positioned as the Special Representative in Europe for *The Australian Women's Weekly* – reported:

> Though it may seem early days to start talking about beach fashions for Australia, summer will soon be round again – and every woman likes to

> be wise to Continental fashions before making advance plans for her warm weather outfits. Beach clothes in Europe this season have been remarkable for their casual, don't-give-a-darn appearance ... Bathing suits have become very daring. Those of which Paris approves most this season are merely composed of brassier and short trunks. And when we say brassier, we mean just that. It is so cut that it not only holds the bust, but gives it form. Another version has a vertical band, enforced by a double fold of the material, which extends from the centre front of the brassier to the waist. This disciplines stomach and diaphragm and helps to hold the figure as slim and flat as it should appear in a bathing suit. (*The Australian Women's Weekly*, 1933a: 2)

Discursive systems operating here mobilise a whole range of commodities by appealing to readers struggling to keep bodies slim at the beach; reproducing understandings of sexually attractive, fashionable feminine bodies. At the same time, the fashionable body challenged moral beach codes through a 'don't-give-a-darn' appearance that exposed for public scrutiny the flesh of stomachs, legs, arms and backs. Nell Murray's report was accompanied by sketches of the 'daring' designs that exposed women's stomachs, shoulders and legs (Figure 6.3).

Nell Murray went on to explain the importance of new fabrics:

> Elastic Costumes! The newest fabric idea consists of the introduction of elastic material for bathing costumes. These fit just like a second skin, and offer support for the figure as well as fulfilling their primary function. (*The Australian Women's Weekly*, 1933a: 2)

Innovations in fabrics allowed the manufacture of swimsuits that did not stretch and sag like those made from wool that clung to dry as well as wet bodies. Jantzen – a knitwear company based in Portland, Oregon, United States of America – marketed 'Miracle Yarn', 'Lastex' and 'Kava Knits'. For example, an advertisement for Jantzen swimsuits in *The Australian Women's Weekly* (Figure 6.4) informed readers that:

> Among the Jantzen presentations for 1935 are two entirely new fabrics – the beautiful Kava-Knits. The luxurious fabrics follow the Continental vogue for novelty surfaced materials and in addition provide the marvellous knitted-in figure control made possible by Jantzen's exclusive process. Never before has such artistry in knitting been combined with such amazing properties of elasticity and resiliency. (*The Australian Women's Weekly*, 1935b: 18)

Figure 6.3 Artist Petrov's ideas of the latest in swimsuits abroad (*Source*: *The Australian Women's Weekly*, 1933a: 2, accessed 29 May 2011. http://nla.gov.au/nla.news-article47471091)

Similarly, in October 1936, an advertisement in *The Australian Women's Weekly* (Figure 6.5) for Jantzen's swimsuits promised:

> a brief, sleek Jantzen that makes you feel like last-minute fashion news from abroad. These new Jantzens have youth and verve and designs that transport genuine Riviera chic to your own favourite beach. But Jantzens also have something more! A *lasting smartness* that is worth all the season's fashion features put together – a *permanent* beauty that no other suit can duplicate. Jantzens remains fresh and attractive and snug-fitting

Figure 6.4 'The new Kava-Knits ... exclusively Jantzen!' (*Source*: *The Australian Women's Weekly*, 1935b: 18, accessed 29 May 2011. http://nla.gov.au/nla.news-article51277226)

through the season because they're made of the world-famous *permanently elastic* Jantzen fabric. (*The Australian Women's Weekly*, 1936a: 48, italics in original)

Advertisements and articles in the *Illawarra Mercury* and *The Australian Women's Weekly* helped to establish and circulate transnational understandings of how to 'dress-up' for the beach. As an advertisement in the *Illawarra*

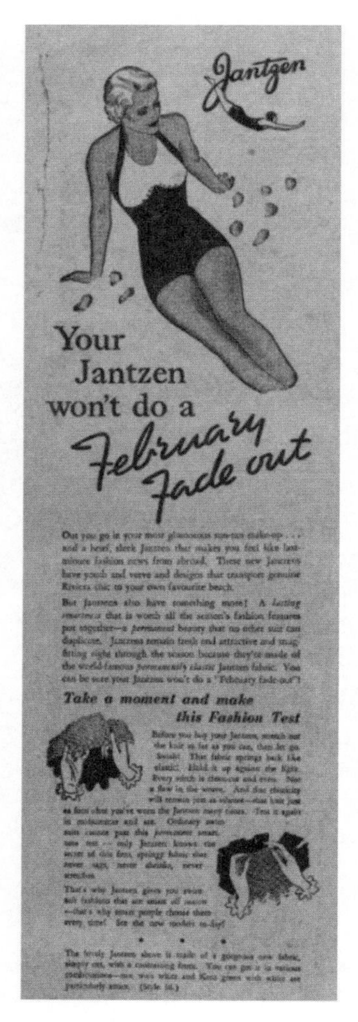

Figure 6.5 'Your Jantzen won't do a February fade out' (*Source*: *The Australian Women's Weekly*, 1936a: 48, accessed 29 May 2011. http://nla.gov.au/nla.news-article47476606)

Mercury in October 1939 from Walter Lance and Co. Pty. Ltd suggested, the modern bather was understood to dress in a different branded bathing costume each year, inspired by what people were wearing to fashionable beach resorts in North America, Britain and Europe (*Illawarra Mercury*, 1939g). The economic function of seasonal swimwear fashions was to make last year's swimwear prematurely obsolete and to accelerate the circulation of goods. For example, an advertisement for Jantzen's swimwear conveyed in block capitals that: 'LANCE'S HAS A COMPLETE RANGE OF 1939–40 JANTZEN' (*Illawarra Mercury*, 1939g).

In addition to the discursive economies within which the leisure and tourism geographies of bathing took shape, advertisements and articles in the *Illawarra Mercury* and *The Australian Women's Weekly* in the 1930s helped imagine a particular version of heterosexuality as moral that did not require the segregation of sexes or the covering-up of bodies from neck-to-knee. Indeed, in December 1933, readers of *The Australian Women's Weekly* were asked the question: 'Is there anything demoralising in it' [the scene (Figure 6.6) depicted on the front page of *The Australian Women's Weekly*, November 25, 1933, 'Sunburn and Laughter...and a Dipping Sea'] (*The Australian Women's Weekly*, 1933c: 12). As a place to discover nature, the beach had particular resonance in marketing swimsuits as an integral part of the construction of heterosexual courtship, romance and romantic love.

Sleeveless, and cut off at mid-thigh rather than the knee, neither costume in this drawing conformed to Council by-laws that imposed neck-to-knee regulations. At the same time, such advertisements helped circulate and naturalise understandings of surf-bathing masculinity as white, able-bodied and athletic, virile and heterosexual. The advertisements of bathers were also a racial mirage generated by a white-dominated society to organise the beach, signalling again how discussion of Aboriginal Australian dispossession were both forgotten and avoided.

Similarly, the marketing of 'His' and 'Hers' swimsuits reinforced heterosexual cultural courtship as appropriate relationships between beach bodies in the 1930s. For example, Figure 6.7 is a copy of a 1937 Jantzen advertisement that appeared in both the *Illawarra Mercury* (1937n) and *The Australian Women's Weekly* (1937a: 36). The Jantzen advertisement used the pitch: 'His Trunks: They're Jantzen's breezy new, long-waisted version of the half-skirted trunk. Narrow belt. Masculine rib stitch'; and 'Her 'bra-lace' gives enchanting grace. A full-skirted suit knitted in another striking new Jantzen fabric.'

Similarly, Figure 6.8 is a reproduction of an advertisement that appeared in *The Australian Women's Weekly* in October 1938 with the slogan: 'Skip the flattery, darling – my Jantzen takes care of that' (*The Australian Women's Weekly*, 1938a: 64). The advertisement went on to read: 'His "HALF-HITCH" trunks' and 'Her "CLIPPER" Suit'. And further, an *Illawarra Mercury* advertisement appeared in October 1939 with the slogan: 'You have a good line darling, but your Jantzen lines are better.' About the male figure it stated: 'He looks Handsome and Manly in sleek, perfectly

THE AUSTRALIAN WOMEN'S WEEKLY

free paper pattern

PRICE 2d. Biggest Value In The World

LARGER CIRCULATION THAN ANY OTHER WEEKLY NEWSPAPER IN AUSTRALIA

Vol. 1. No. 25. SATURDAY, NOVEMBER 25, 1933. Phones: M 2081 (4 lines). 48 PAGES

SEASIDE Holiday

Verse by . . . P. DUNCAN BROWN

"Sunburn and Laughter . . . and a Dipping Sea"

WYNNE W. DAVIES

Come away . . .
Come, in the Summer madness,
Come with your youth and gladness,
Forget night is coming after,
Live for the moment,
Live for the laughter
Of to-day.

To the sea
Where the curling combers wake,
To foam fretted patterns break . . .
Plunging into the greeny depths
With your open eyes,
With your bated breaths . . .
Ecstasy . . .

Come away . . .
The tides are swelling full and keen . . .
Limbs gleam golden under the green,
The splashing, tumbling hours flee,
Sunburn and laughter
And a dipping sea . . .
Holiday.

Figure 6.6 'Seaside Holiday: 'Sunburn and Laughter . . . and a Dipping Sea'' (*Source*: *The Australian Women's Weekly*, 1933b: 1, accessed 29 May 2011. http://nla.gov.au/nla.news-article48204747)

fitting trunks in the new 'Suede-Sheen' fabric' (*Illawarra Mercury*, 1939g). As argued by Stevi Jackson (1999: 103), one consequence of romantic heterosexual love is to 'validate sexual activity morally, aesthetically and emotionally'. Lynda Johnston and Robyn Longhurst (2010: 134) pointed out that another consequence of discourses of romantic heterosexual love is to 'alleviate fears about sexual and emotional exploitation'. Silenced by the excitement of romantic love's passion are the uneven social relationships of heteropatriarchy. The beach is not a passive backdrop in framing romantic heterosexual love. Instead, how the beach was framed as an

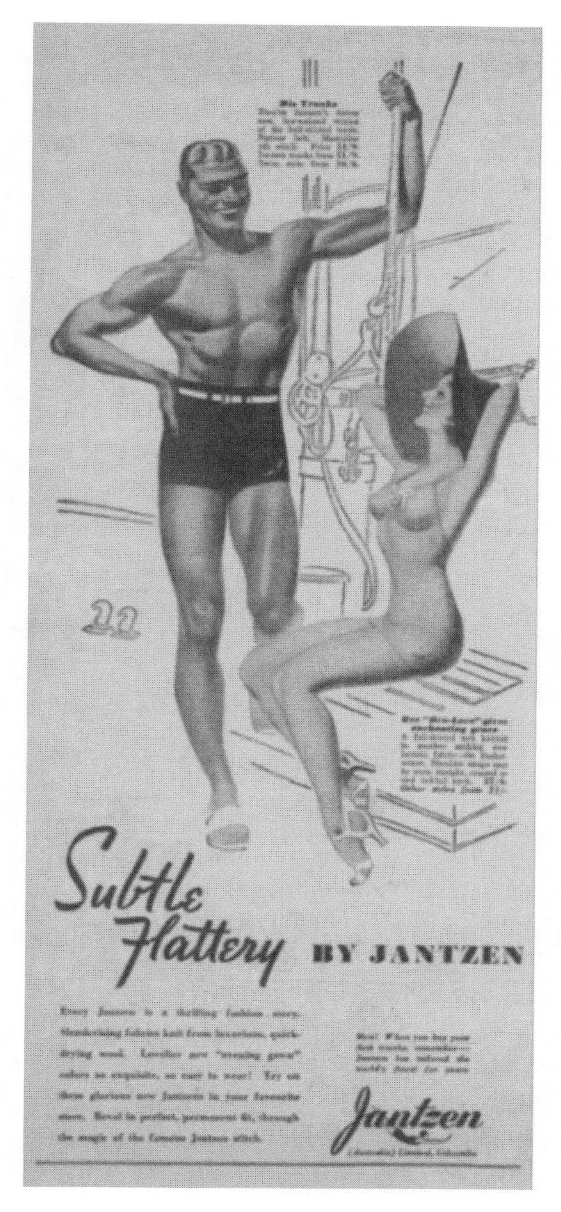

Figure 6.7 'Subtle Flattery by Jantzen' (*Source*: *The Australian Women's Weekly*, 1937a: 36, accessed 29 May 2011. http://nla.gov.au/nla.news-article52256310)

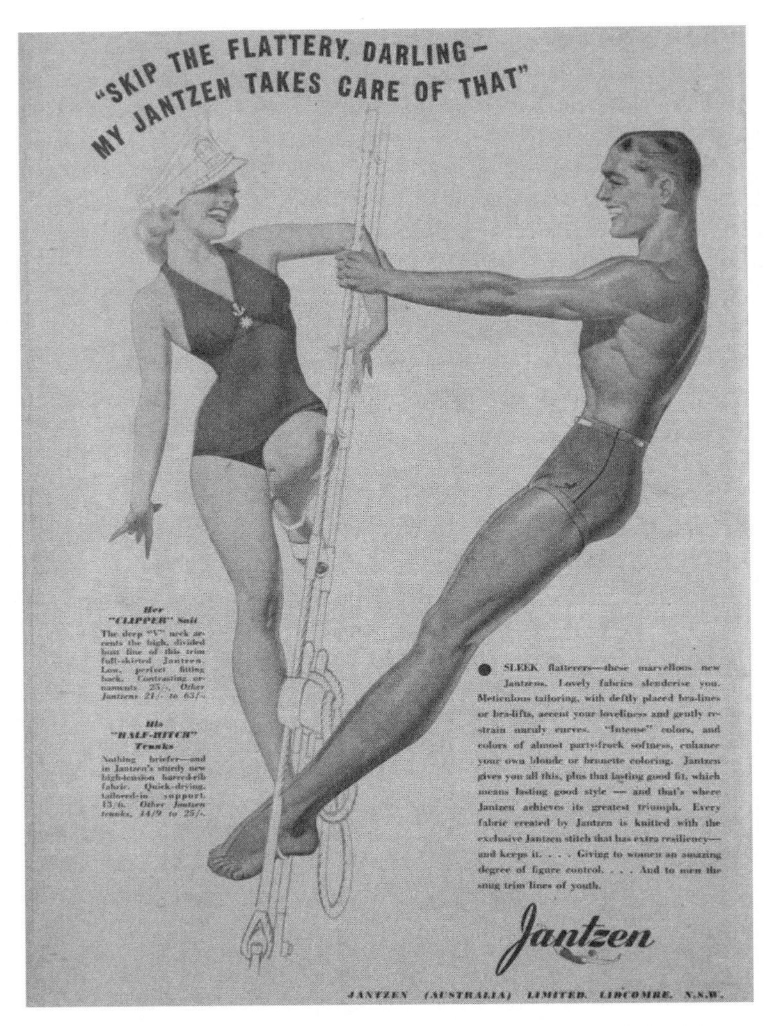

Figure 6.8 'Skip the flattery, darling – my Jantzen takes care of that' (*Source*: *The Australian Women's Weekly*, 1938a: 64, accessed 29 May 2011. http://nla.gov.au/nla.news-article52256310)

earthly paradise, helped sustain flows of single heterosexual bodies to the beach, where courtship could be understood as romantic, pure and natural. Heterosexual courtship in public was naturalised for some at the beach in the 1930s.

Swimsuit Bodies

The marketing of swimming costumes provided insights into how what was regarded as heterosexually attractive bodies in the 1930s was framed by discourses of the physical culture of fitness. Jantzen swimsuits helped to stabilise the shape and size of sexed, gendered, racialised and fit bodies that were culturally valorised at the beach. Swimsuits were marketed towards beach-goers who wanted to retake control of his or her body. Women were offered:

> Lovely fabrics slenderise you. Meticulous tailoring, with deftly placed bra-lines or bra-lifts, accent your loveliness and gently restrain unruly curves ... Giving to women an amazing degree of figure control. (*The Australian Women's Weekly*, 1938a: 64)

In the late 1930s, priority was given to shaping a women's body to ensure it was sculpted and streamlined, but emphasised the breasts. The mastery of the body, in this case, relied upon the scientific discourses of new fabrics rather than exercise. The swimsuit was positioned as a source of empowerment for women because it promised to keep the 'unruly curves' of the female body in 'proper' shape. Set firmly within the traditional Western epistemology, female flesh was positioned as a problem and technology of swimwear was offered as a solution.

In contrast, Janzten swimwear satisfied the needs of men who wanted 'the snug trim lines of youth' (*The Australian Women's Weekly*, 1938a: 64). The advertisement for Jantzen swimwear in the *Illawarra Mercury* in December 1937 (*Illawarra Mercury*, 1937n) and *The Australian Women's Weekly* (1937a: 36) point towards how trunks were now fashionable for the embodiment of normative heterosexual masculinity at the beach. It was young men who went topless at the Australian beach to emphasise their difference and otherness to women.

The normative masculine body emphasised the chest and shoulders; they were required to be young, hairless, bronzed, muscular and sculpted. As illustrated in Figures 6.7 and 6.8, Janzten's figuration of control and solidity for men appropriated myths that appealed to the bronzed, well-sculpted, lean, muscular masculinity. The image of the topless male swimmer was culturally intelligible because it reproduced the masculine/feminine gender dualism. Furthermore, as outlined by Douglas Booth (2001) the figures represented in the Janzten advertisements, were located not only within particular cultural constellations of fitness, eugenics, sexuality and gender, but also social class, age and race. Hence, the position of the surf-bather was therefore always controversial. The next section explores this controversial positioning of the heterosexual surf-bather at the beach. Some people, categorised in the media as 'Mrs Grundy' for their conventional values, constituted the heterosexual bodies of surf-bathers clad in trunks and backless bathing costumes at the beach as immoral and disgusting, rather than fashionable and sexy. Boundaries between moral and immoral heterosexual identities were created and challenged in and through the social relationships that comprised the Australian beach in the 1930s.

Shrinking bathing costume fashions were controversial up to the late 1930s in Australia. Swimwear fashions for men and women in the 1930s broke the neck-to-knee rules governing bathing costumes. Readers of *The Australian Women's Weekly* were reminded about the laws that prohibited wearing trunks and backless costumes in October 1933 (*The Australian Women's Weekly*, 1933d: 4), and again in July 1934 (*The Australian Women's Weekly*, 1934b: 4). As noted in these articles, following the Local Government Act of 1919, Councils in New South Wales had 'the last say as to what is or is not an indecent costume' (*The Australian Women's Weekly*, 1933d: 4). These articles revealed differences between how Councils in Sydney interpreted and enforced the bathing ordinance. Waverly Council was positioned as 'very progressive'. In contrast, it was pointed out that Cronulla Council enforced men to wear shorts over their costumes when not in the water. One article warned that: 'The Mrs Grundys of these councils are as shocked at bare limbs and bare backs as their grandmothers of Queen Victoria's day' (*The Australian Women's Weekly*, 1933d: 4).

In the 1930s, it was topless men at the beach that created moral narratives around heterosexuality. The boundaries between moral and immoral heterosexual identities were created, sustained and challenged by bodies dressed in trunks at the beach. Letters to the editor of *The Australian Women's Weekly* illustrated how understandings of heterosexuality at the beach were entangled into gender, age, race and age. For example, D.A. Jurns wrote:

> some men have no shame at all, and I think there is nothing more horrid than to see a man's hairy chest. Surely they could arrange something to hide this. Why can't they have their bathers cut similar to the ladies', high in the front and low at the back. Let's hope for this anyway. (*The Australian Women's Weekly*, 1933e: 17)

For D.A. Jurns, bodies dressed in trunks at the beach were horrid. As argued by Iris Young (1990) bodies constituted as ugly are those that operated outside of sexual moral norms. D.A. Jurns also tapped into moral narratives of shame to put forward the idea of surf-bathers' bodies dressed in trunks at the beach as not fitting within their understanding of respectable heteromasculinities. Similarly, Mrs E.G. Woodger, wrote:

> I sincerely hope that we shall never see men parading our beaches in trunks. I think it would be a most repulsive site ... I do not think that a half nude male with his hairy chest is an object for admiration. (*The Australian Women's Weekly*, 1934c: 13)

Like D.A. Jurns, Mrs E.G. Woodger also sought to exclude bodies dressed in trunks from the beach. Hairy chests and the sexualisation of the surf-bathing body posed a threat to the moral norms of the beach. Indeed, Kay Hearfield asked the question in her letter to the editor: 'What girl would want to romp and play on the beach with a man who was naked to the waist?' (*The Australian Women's Weekly*, 1934d: 13). For Kay Hearfield, the semi-naked bodies of men in public bathing reserves engendered moral decay. For opponents, being dressed in trunks tended to fall outside conventional gender, sexual and moral norms.

In contrast, advocates for men-in-trunks employed discourses of rational science, Greek civilisation and the physical culture of swimming. For example, C. Halbert argued that:

> Considered biologically, the hairy chest is a manifestation of virility and essential masculinity; moreover, it is a symbol of sex differentiation. (*The Australian Women's Weekly*, 1933f: 17)

While, Miss J. Gilmore pointed out:

> The Greeks who possessed an older and some historians say, a far higher form of civilisation than ours, believed that the acme of civilisation would be reached when men could go about their day's work unclothed and entirely unconscious of the appearance of their bodies. If this theory is correct, we of the modern world are still in a barbarous state. (*The Australian Women's Weekly*, 1933g: 17)

Mrs G.K. Yorkston wrote:

> I do not see any reason at all why men should not wear trunks alone. When a man goes on the beach he goes for a sunbath and a swim. How is one to have a sunbath hampered with a costume? And, isn't it far better when swimming to have on as little as possible? Some women,

> perhaps, would think them repulsive, but there are always a certain few opposed to anything modern. (*The Australian Women's Weekly*, 1934e: 13)

Finally, Mrs W. Foster wrote:

> There is nothing either repulsive or ludicrous about a muscular male torso provided that it is brown. (*The Australian Women's Weekly*, 1934f: 13)

However, the economic discourses of fashion and the rational discourses surrounding sunbathing, swimming and virility in forging the modern male subject were trumped within the Legislative Assembly of New South Wales by discourses that positioned semi-naked surf-bathing bodies outside of sexual and moral norms. In 1935, Eric Spooner, the Minister for Local Government, successfully amended the Local Government Ordinance No. 52 impacting bathing costumes. Under the new Ordinance all bathing costumes, for a person over four years of age, on a public bathing reserve must cover the whole of the front of the body below the level of the armpits, the whole of the back and cover at least three inches of the legs from the waist line. And for persons over 12 years of age, a skirt, or a front half-skirt was made compulsory from the waistline to the lower end of the leg covering. The underlying imperative of this legislation was to cover up genitals that sexualised bodies. Spooner's bathing legislation drew on understandings of sexed bodies in public as immoral, as discussed in Chapter 2.

Nevertheless, the archival records from the Illawarra suggested that reporters ridiculed the chance of Councils enforcing Spooner's amendment to the bathing ordinance. One anonymous reporter in the *Illawarra Mercury* joked that newly appointed beach inspectors were given rulers to enforce the prescribed neck-to-knee costume:

> About ten members were elected as beach inspectors, to be approved by the Council . . . They tell me that they all have rulers to measure Mr. Spooner's costume, to see that 3in. leg is the fashion. (*Illawarra Mercury*, 1935k)

Within local councils there was also evidence of councilmen being divided over the moral and immoral boundaries of the surf-bathers' bodies dressed in trunks. In 1937, the *Illawarra Mercury* reported on 'important discussions' and 'much-discussed questions' within the Bulli Shire Council meeting caused by a Mr Woodhill, who was 'swimming and sunbaking in trunks' at Austinmer beach (*Illawarra Mercury*, 1937o). Bulli Shire Council was divided whether to take legal proceedings against Mr Woodhill after he had refused to follow the beach inspector's instructions to either leave the beach or dress in a regulation bathing costume. Cr. Clowes recognised that if the Council did not take action:

> it will mean the regulations will not be of any avail in the future, while a prosecution would serve to show Council is out to put the regulations into effect.

Cr. Fackender agreed with Cr. Clowes saying: 'the Inspector was appointed for the special purpose, and Council should back him up'. However, Cr. Quilkey is quoted as stating that he considered bathing in trunks to be 'quite decent', and was 'satisfied those wearing trunks to-day were paving the way for what will happen next surfing season'. As argued by Booth (2001: 48), Cr. Quilkey's position reflected 'the failure of Mrs Grundy to prove that exposed flesh engendered moral decay'. Instead, semi-naked masculine beach bodies were being shaped by discursive systems of health, eugenics, gender, race and sexuality that produced 'good' national citizens, embodied in the surf lifesaver (see Chapter 5). After many amendments, that were not carried, Council decided that Mr Woodhill would be 'asked to apologise, failing which a prosecution to follow' (*Illawarra Mercury*, 1937o). Councillors in the Illawarra were not prepared to act unanimously to enforce regulation over the wearing of 'trunks only'. Indeed, Cr. Clowes, who wished to take action against men wearing 'trunks only' at the beach in 1937, appeared to be supportive of men wearing 'underpants' at the beach only a year later:

> Cr. Clowes replied that probably the underpants were equally as sufficient covering as are some bathing costumes worn on the beaches. (*Illawarra Mercury*, 1938i)

Cr. Clowes' shift is just one example of the rapidly changing ideas in New South Wales towards the display of semi-naked bodies at the beach understood as masculine.

Tanning Bodies

> It would not be an exaggeration to say, Europe has gone sun-tan crazy, or that the average smart English woman would give a great deal to possess the natural sun-tan complexions of her sisters in Australia. (*The Australian Women's Weekly*, 1934g: 37)

Tanning bodies on Australian beaches has received substantial attention in the literature (Booth, 2001; McGloin, 2005; Ramsland, 2000; Rodwell, 1999; White, 2007). Most of this work has examined the role of tanning in stabilising gendered and racialised subjectivities. As argued by Booth (2001), during the first decades of the 20th century Surf Life Saving Associations (SLSAs) were integral to positioning the tanned and gendered body at the beach as integral to fixing understandings of the national scale. Tanned and gendered bodies at the beach were one way the SLSAs made visible the Australian nation. Visibility of the nation being performed through sun-basking at the beach required drawing on the gendered and racialised discursive framings or 'truths' of ancient Greek civilisation, eugenics and the medical science of heliotherapy (see Chapter 2). Tanning bodies at the beach illustrated the importance of the politics of geographical scale and how the constructions of the national scale through tanned bodies by the SLSA were deployed to legitimise surf-bathing as an exclusively masculine activity. Tanning bodies on the beach highlight Doreen Massey's (2005) and Neil Smith's (1992) argument that geographical scale, when conceptualised as a practice, rather than as an absolute category, can provide insights into processes of social inclusion and exclusion. The SLSA created the expectations that Australian bodies were to be tanned and 'glowing'. As argued by Sara Ahmed (1997) through tanning practices these discursive truths about the nation were 'written' on the skin.

Tanning bodies at the beach also became a way of practising 'modern' summer leisure and tourism. Sun-basking on the sand became popular at Australian beaches among followers of European fashion trends that framed tanned skin in terms of markers of youth, leisure, wealth, health and female beauty. As argued by Fred Inglis (2000), in Britain during the 1930s, the tanned body suggested having the wealth of both leisure and overseas travel to Mediterranean resorts. No longer were bourgeois bodies in Europe marked by their pale skin. Instead, the tan and sun-basking became associated with leisured female beauty of fashionable beach resorts of the Mediterranean (Figure 6.9).

However, in the racialised context of Australia, where identity was often defined in terms of the colour of the skin, the sun-tanned body risked losing being associated with the white colonial culture. Responding to an article discussing the popularity of sun-baking in England in 1933, Miss S. Hyde, tapped into racialised discourses surrounding skin as a hegemonic indicator of the 'truth' of a subject to highlight how tanning practices took on different meaning in Australia:

> Confusing for Savages ... One of the earliest steps taken to stimulate the self-respect of the 'abject half-caste' is to provide her with clothes.

BARE SKIN *Craze*
Sweeps ENGLAND

OUR LONDON REPRESENTATIVE declares that girls have gone further than ever before this year in the wearing of bathing costumes, shorts, and other scanty apparel. This photo. shows a holiday maker assisting a Guildford farmer gathering the harvest.

Law Favors Sunbathers In Preference to Mrs. Grundy

Figure 6.9 'Bare Skin *Craze Sweeps* England' (*Source*: *The Australian Women's Weekly*, 1933i: 2, accessed 29 May 2011. http://nla.gov.au/nla.news-article48203897)

> She whose forebears have roamed beneath the sun in unclothed simplicity for generations, is given an all enveloping frock. On the other hand, the English girl, clothed to the eyebrows for centuries, has suddenly discovered the sun, and to celebrate goes all native in brassiere and shorts. (*The Australian Women's Weekly*, 1933h: 14)

Miss S. Hyde drew on racist stereotypes associated with Aboriginal Australians to question sun-bathing practices in England in 1933. In the colonial spaces of Australia, skin was still understood as marking the boundary between the bodies of 'savages' and 'civilisation'. As Ahmed (1998) argued skin can be conceptualised as a border between 'self' and 'other'. In Australia, during the 1930s, tanned skin risked losing its civilized embellishment status by being read and essentialised as Aboriginal.

However, articles written for *The Australian Women's Weekly* emphasised the importance of being tanned as an expression of sexually attractive, youthful, 'white' femininity. Tanning bodies are framed through the discourse of beauty and leisure. For example, Miss Kathleen Court, 'the well-known cosmetician who had just returned from a long tour of England and the Continent', informed readers of the 'New trends in Beauty Fashions' and how: 'English women covet the deep tan which the Australian sun imparts to our women, and are always striving to acquire it' (*The Australian Women's Weekly*, 1934h: 22). Kathleen Court draws on privileged nationalistic discourses of whiteness to naturalise tanning and what it means to be an Australian.

Throughout the 1930s readers of *The Australian Women's Weekly* were given annual advice on how to enjoy their leisure and beach holidays by acquiring a suntan. For example, beauty editor Evelyn advised readers of *The Australian Women's Weekly* in September 1934 that:

> you must have the golden, sun-tanned look, which is everywhere the vogue, and a mastery of the new sun-tan make-up will enable you to acquire it, and at the same time to protect your skin from the damaging effects of too much sunshine. (*The Australian Women's Weekly*, 1934g: 38)

As Ahmed (1998) argued, the skin is a site of contradictions. Tanning practices on the sands simultaneously enhanced and destroyed the feminised subject. Tanned bodies to become more feminine, needed to be attentive to avoiding how tanning also is the cause of burns, wrinkles, freckles, lines, premature ageing and cancer. These contradictions were again explicit in 1935 by Evelyn, the beauty columnist. She wrote:

> Remember that while the sun is life-giving, it is also death-dealing when taken in excess – death-dealing to delicate skin and feminine beauty. People recklessly bathe themselves in sunshine, and only when too late – when faces are red and sore and stiff, skin coarsened and thickened – do they realise they have been doing themselves active harm. (*The Australian Women's Weekly*, 1935c: 51)

In the 1930s, tanned skin was positioned paradoxically as both a marker of healthy and damaged skin; enhancing and detracting femininity. In the 1930s, *The Australian Women's Weekly*, drew on both medical research and the cosmetics industry to circulate a discourse of 'wise' tanning. Readers were provided with rules of how to tan by limiting their exposure to sunlight, thereby enhancing rather than diminishing leisured female beauty. Warnings of the potentially harmful effects of tanning and rules for 'wiser' tanning were outlined in the following articles: 'The Body Beautiful. When you go down to the sea, don't overcook yourself, fair maid, in a hurried, futile effort to acquire bronze loveliness' (*The Australian Women's Weekly*, 1936b: 75); and 'Outdoor Loveliness For You! How to acquire a nice tan with no freckles or burns to spoil the effect of delicate party frocks' (*The Australian Women's Weekly*, 1937b: 63). Bodies to appear feminine through sun tanning were advised to follow strict pre- and post-beauty regimes of oiling, moisturising and cleansing to minimise skin damage. In 1938, the medical correspondent for *The Australian Women's Weekly* warned of skin cancer from tanning practices, and provided the following strict advice: 'Brunettes – Go Ahead. Blondes – be Careful. Redheads – Refrain!' (*The Australian Women's Weekly*, 1938b: 4). In the late 1930s, tanning was subject to counter discourses from medical organisations. However, sunburn and skin cancer were interpreted as holding essentialised biological 'truth' about race. For example, an article written 'by a doctor' for *The Australian Women's Weekly* in January 1940, concluded 'as a white race our skins are not perfectly adapted for exposure to the sun, and we must use our reputed to be superior intelligence in adapting ourselves to our sub-tropical environment' (*The Australian Women's Weekly*, 1940: 45). Tanning rules were positioned by this 'doctor' as further evidence of a fixed racial identity, and the strict racialised boundaries between British and Aboriginal Australians.

Conclusion

This chapter presented five examples of how bodies at the beach were framed as 'modern' in the Illawarra in the first decades of the 20th century. Discourses of nature, science, fashion, cosmopolitanism and romantic love were deployed to make bathing modern. These examples illustrate how bodies cannot be separated from space and the multiplicities of heterosexual cultures. The beach is an important site for understanding the intersection between sexuality, gender, class and race. The beach is also forged through the interplay of connections, movements and intersections across different spatial scales. Furthermore, the production of the beach in Australia is always a political process.

The controversy over whether gendered bodies should be allowed to bathe together in public at the beach and how swimming bodies should be dressed at the beach illustrated that there is nothing inherent in the beach that makes particular expressions of heterosexuality more or less moral or appropriate. Instead, it is the intersection of different discursive framings about bathing, swimming, and swimsuits that fashioned whether those bodies belonged at the beach or were positioned as sexy, masculine, feminine, heroic, disgusting, indecent or shameful.

This chapter also illustrated how the politics of scale were deployed as a strategy of empowerment as well as to dominate, control and define others. For example, Eric Spooner's amendments of the ordinance in 1935 were a clear illustration of how through reconfiguring the boundaries of the bathing reserve those not dressed in neck-to-knee bathing costumes would be configured as indecent. In contrast, fashion reporters like Nell Murray for *The Australian Women's Weekly* deployed the international scale as a strategy to help young consumers break free from the constraints of the New South Wales bathing reserve politics by providing access to trends in the French Riviera, London and Hollywood. At the beach, the politics of scale was employed as both a form of transformation and control in the first decades of the 20th century.

Finally, regardless of the particular expression of heterosexual culture, the chapter illustrated the continuity of the power associated with masculinity in configuring the spaces of the beach. One example is how some women complained of 'pirate' pests, and the normative assumption that all women at the beach could be 'chatted up'. Another example of how the beach was configured as a masculine space was illustrated through the marketing of bathing costumes. White masculinity was sexualised through naturalised discourses of virility and well-toned physiques. White femininity was sexualised through pitching bathing costumes that offered greater 'loveliness' and control of the body. Furthermore, one of the enduring myths deployed by swimsuit marketers was the romantic love narrative with 'his' and 'her' swimsuits. For habituated-followers of fashion, the commercialisation of beach culture offered empowerment by transcending the regulation of the bathing reserve, but, layered another mechanism of regulation or discipline over bodies at the beach through a market that established normative ideas of sexuality, glamour and romantic love. The pitch of romantic heterosexual love silenced the discursive system that produced and regulated the uneven sexed identities of men and women at the beach.

Conclusion

Just as none of us is outside or beyond geography, none of us is completely free from the struggle over geography. That struggle is complex and interesting because it is not only about soldiers and cannons but also about ideas, about forms, about images and imaginings
Said, 1993: 6

This is a book that engages with the struggles over the geographies of the Australian beach: its makings, boundaries and meanings for the West. *Tourism and Australian Beach Cultures: Revealing Bodies* is focused on the geographical form of the beach. The beach is an important and multivalent place in Australia. As Fiske *et al.* (1987: ix) argued the beach is a site 'where Australians construct (and deconstruct) a plenitude of meanings, using a multitude of practices – not a single meaning with a single value'. These authors along with Douglas Booth (2001) and Ed Jaggard (1997) have demonstrated the 'richness' of meaning in Australian beach cultures, rather than a 'single truth'. Furthermore, cultural studies have noted how the idea of 'the beach' is mobilised as a site of foundational myths about the ordinariness of Anglo-whiteness. Taking a cue from Meaghan Morris (1992, 1998), Bonner *et al.* (2001: 270) noted how the beach was found to be a 'privileged site' for the exploration of Australian national identity in cultural studies. John Hartley and Joshua Green (2006: 348) noted that Morris identified the beach 'as an often-compelling object utilized by Australian cultural studies to develop its project, full of excess signification for Australian cultural studies to reveal'. Morris (1992, 1998) warns of starting with the idea of the beach as the supposed terrestrial solidity that founded the imaginative geography of Australia.

The response of this book to such challenges was to explore the beach as a constellation of trajectories, and a spatialised 'geometry of power' for unpacking geopolitical-embodied effects and ideologies that are precariously shifting, vulnerable and unfolding (Massey, 2005). The aim of *Tourism and Australian Beach Cultures: Revealing Bodies* was to provide a spatial interpretation of the historical relationship between bodies, beach cultures and tourism in Illawarra, New South Wales. The beach was selected as a site where sexed, gendered, classed and racialised identities could be explored. The focus of *Tourism and Australian Beach Cultures: Revealing Bodies* was the 19th and early 20th century. The period 1830–1940 was chosen to explore the ideas circulating about bodies at the medical and pleasure beach – either clothed or unclothed – as both imperial and national projects.

An historical archive was comprised of records from newspapers, magazines, the Bank of New South Wales, the New South Wales Government Railways and surf clubs. Assembling and interpreting these records exposed an obvious silence of Indigenous voices, along with women and the working classes. These were clearly limitations. The archive provided a partial insight predominantly comprised of mainstream British or Anglo-Australian, male, middle-class narratives. However, despite these limitations, the archive remained an important source of material to analyse how mainstream narratives and counter narratives operated to reveal bodies as sexed, gendered, classed and racialised at the beaches of the Illawarra. The compiled historical archive from the then out-of-the-way beaches of Illawarra, New South Wales was crucial to upset the general metropolitan priority focus on beach cultures. Sydney is often given primacy in historical discussion of the politics and representations of the beach. This compiled archive provided historical and ideological context by which the Illawarra beach was shaped and reshaped.

Tourism and Australian Beach Cultures: Revealing Bodies is enriched by drawing on feminist post-structuralist theory to explore the reciprocal relationships that simultaneously constitute bodies and spaces of the beach. The beach is argued to have shaped bodies as much as bodies shaped the beach. Equally, our argument relied upon uncovering the myriad connections across geographical scales from the body, region, nation and empire. This conceptual apparatus enabled us to explore what it meant to 'perform' different subjectivities in and through the beach.

Tourism and Australian Beach Cultures: Revealing Bodies is a small step that helps to address the lack of critical historical analysis into the geographical imaginary and practices of beach resorts. Furthermore, most scholarly discussions of the Australian beach in history, cultural studies and sociology, while emphasising race, class and gender, the bodies' genitalia seem not to have been discussed.

Premised to the fiction of a place untouched by outside, *terra nullius*, the geographies of the Illawarra beaches were constituted by racial difference. Central to the argument was to illustrate the importance of recognising the importance of sexed bodies in and through bathing practices at the beach. Through our entry point of the geographical scale of the body we illustrated how different leisure practices at the beach became a spatial, moral and cultural boundary which divided 'self' and 'other', 'civilised' and 'primitive', 'masculine' and 'feminine', 'art' and 'obscenity'.

Tourism and Australian Beach Cultures: Revealing Bodies traces the history of the conjunction of ideas, images and imaginings in the formation of the geographical form of the beach through key episodes that 'disciplined' sexual bodies. The beach is not exempt from the space of history. The historical weight of discourses that fix the bounded territoriality of the beach cannot be overlooked. The first chapter described how naked bathing bodies in public were constituted as obscene because of how the body was sexualised. Unlike in Britain's fashionable bathing resorts, daylight bathing in public view was banned at bathing resorts in New South Wales following the passage of an 1838 Act. The nude male bathing body at the beach became a key marker of the border between moral and immoral heteromasculinities. The second chapter outlined the new spatial order of the bathing reserve. In New South Wales the passage of the Local Government Act of 1906 established the bathing reserve, facilitating the public appearance of the sanctioned neck-to-knee bathing costume. The bathing reserve enabled the excision and isolation of the beach from surrounding places. Bathing ordinances protected and regulated the margins that became the beach. Bathing ordinances mapped the limits of the beach and defined what actions were acceptable to the Councils. These chapters provide the backdrop for the discussion of defining events in the first decades of the 20th century. These ranged from the arrival of mass tourism facilitated by the steam train, elite day-trippers propelled by the automobile, surf lifesavers and images of Hollywood movie stars and international beach fashions. Having drawn lines that constituted the moral and spatial boundaries of the beach, these chapters explored how individuals negotiated the formal prohibitions against the display of the sexualised body at the beach. Within the boundaries set by the formal law, understanding the beach required turning to how different social groups made their own lore, their own space and their own sense of self through embodied practice. *Tourism and Australian Beach Cultures: Revealing Bodies*, suggests geographies of the beach are less about formal laws, but about how different beach users like surf lifesavers, swimmers and sunbathers, generated narratives and actively carved out spaces at the beach through embodied knowledge and repeated practice. Some social groups went to the beach to actively seek pleasure by flirting, while others sought the bonds of mateship. Bodies at the beach were always political through how they enacted geographies of connection and disconnection to carve out spaces through a process of negotiation.

Tourism and Australian Beach Cultures: Revealing Bodies, therefore, further opens up tourism studies and tourism geographies to examine the politics of embodiment and the sexed, sexy and sensual body at the beach. Important steps in disrupting the rationalism of the disciplines of tourism studies and tourism geographies of the beach are already illustrated in the work of Caroline Daley (2005); Susan Frohlick (2007, 2008); Lynda Johnston (2005b); Linda Malam (2004); Annabelle Mooney (2005); and Joan Phillips (2002). Continued investigation is required of the ways in which power and knowledge are produced, reproduced and challenged through sexed, sexy, sensual and sexual bodies at the beach.

References

Ahmed, S. (1997) It's a sun-tan, isn't it?: Autobiography as an identification practice. In H. Mirza (ed.) *Black British Feminism* (pp. 153–167). London: Routledge.

Ahmed, S. (1998) Animated borders: Skin, colour and tanning. In M. Shildrick and J. Price (eds) *Vital Signs: Feminist Reconfigurations of the Biological Body* (pp. 45–65). Edinburgh: Edinburgh University Press.

Allen, J. (1990) *Sex and Secrets: Crimes Involving Australian Women Since 1880*. Melbourne: Oxford University Press.

Anderson, B. (1983) *Imagined Communities: Reflections on the Origins and Spread of Nationalism*. London: Verso.

Bashford, A. (2004) *Imperial Hygiene: A Critical History of Colonialism, Nationalism and Public Health*. London: Palgrave Mamcillan.

Bashford, A. and Hooker, C. (2001) Introduction: Contagion, modernity and postmodernity. In A. Bashford and C. Hooker (eds) *Contagion: Historical and Cultural Studies* (pp. 1–12). London: Routledge.

Bayley, W. (1960) *Kiama Municipality*. Notes by William A. Bayley. Vol. 2, 1891–1960, unpublished manuscript.

Bayley, W. (1975) *Black Diamonds: History of Bulli District, New South Wales*. Bulli, New South Wales: Austrail Publications.

Bell, D. (2006) Variations on the rural idyll. In P. Cloke, T. Marsden and P. Mooney (eds) *Handbook of Rural Studies* (pp. 149–160). London: Sage.

Bell, D., Binnie, J., Cream, J. and Valentine, G. (1994) All hyped up and no place to go. *Gender Place and Culture: A Journal of Feminist Geography* 1 (1), 31–47.

Bell, D. and Valentine, G. (1995) Introduction. In D. Bell and G. Valentine (eds) *Mapping Desire: Geographies of Sexualities* (pp. 1–27). London: Routledge.

Bell, D. and Holliday, R. (2000) Naked as nature intended. *Body and Society* 6 (3–4), 127–140.

Beresford, Q. and Omaji, P. (1998) *Our State of Mind: Racial Planning and the Stolen Generations*. Fremantle, WA: Fremantle Arts Centre Press.

Blunt, A. (1994) *Travel, Gender, and Imperialism: Mary Kingsley and West Africa*. New York: The Guilford Press.

Bonner, F., McKay, S. and McKee, A. (2001) On the beach. *Continuum: Journal of Media and Cultural Studies* 15 (3), 269–274.

Booth, D. (1991) War off water: The Australian Surf Life Saving Association and the beach. *Sporting Traditions* 7 (2), 134–162.

Booth, D. (1997) Nudes in the sand and perverts in the dunes. *Journal of Australian Studies* 21 (53), 170–182.

Booth, D. (2001) *Australian Beach Cultures: The History of Sun, Sand and Surf*. London: University of Otago.
Booth, D. (2006) Sites of truth or metaphors of power? Refiguring the archive. *Sport in History* 26 (1), 91–109.
Boscagli, M. (1996) *Eye on the Flesh: Fashions of Masculinity in the Early Twentieth Century*. Boulder, CO: Westview Press.
Bourdieu, P. (1977) *Outline of a Theory of Practice*. Cambridge: Cambridge University Press.
Bourdieu, P. (2000) *Pascalian Meditations* (R. Nice, trans.). Oxford: Polity Press in association with Blackwell.
Browne, K. (2009) Women's separatist spaces: Rethinking spaces of difference and exclusion. *Transactions of the Institute of British Geographers* 34 (4), 541–556.
Bryant, E. (1981) *Beaches of the Illawarra*. Wollongong Studies in Geography, No 5. Wollongong University Geography Department.
Burke, E. (1757) *A Philosophical Enquiry into the Origin of our Ideas of the Sublime and Beautiful*. London: R. and J. Dodsley.
Butler, J. (1990) *Gender Trouble: Feminism and the Subversion of Identity*. New York: Routledge.
Butler, J. (1993) *Bodies that Matter: On the Discursive Limits of "Sex"*. New York: Routledge.
Butler, J. (1997) *Excitable Speech: A Politics of the Performative*. New York: Routledge.
Byrne, D. and Nugent, M. (2004) *Mapping Attachment: A Spatial Approach to Aboriginal Post-Contact Heritage*. NSW Department of Environment and Conservation, Hurstville, NSW.
Castle, R. (1997) Steel city: The economy, 1945–1995. In J. Hagan and A. Wells (eds) *A History of Wollongong* (pp. 71–80). Wollongong: The University of Wollongong Press.
Clark, C. (1978) *A History of Australia IV: The Earth Abideth Forever, 1851–1888*. Melbourne: Melbourne University Press.
Clarsen, G. (2008) *Eat my Dust: Early Women Motorists*. Baltimore, MD: John Hopkins University Press.
Cogdell, C. (2003) Products or bodies? Streamline design and eugenics as applied biology. *Design Issues* 19 (1), 36–53.
Cousins, A. (1994) *The Garden of New South Wales: A History of the Illawarra and Shoalhaven Districts, 1770–1900*. Wollongong, New South Wales: Illawarra Historical Society.
Cover, R. (2003) The naked subject: Nudity, context and sexualization in contemporary culture. *Body and Society* 9 (3), 53–72.
Cumes, J. (1979) *Their Chastity was Not too Rigid: Leisure Times in Early Australia*. Melbourne: Longman.
Daley, C. (2003) *Leisure and Pleasure: Reshaping and Revealing the New Zealand Body 1900–1960*. Auckland: Auckland University Press.
Daley, C. (2005) From bush to beach: Nudism in Australasia. *Journal of Historical Geography* 31 (1), 149–167.
Davidson, J. and Spearritt, P. (2000) *Holiday Business: Tourism in Australia since 1870*. Carlton, Victoria: Miegunyah Press.
Dietrich, H. (1913) Heliotherapy with special reference to the work of Dr. Rollier at Leysin. *The Journal of the American Medical Association* 61 (25), 2229–2232.
Douglas, M. (1966) *Purity and Danger: An Analysis of the Concepts of Pollution and Taboo*. London: Routledge.
Doyle, D.A. (2005) The salt water washes away all impropriety: Mass culture and the middle-class body on the beach in turn-of-the-century Atlantic City. In L. Dowler, J. Carubia and B. Szczygiel (eds) *Gender and Landscapes: Renegotiating Morality and Space* (pp. 94–108). London: Routledge.

Dutton, G. (1985) *Sun, Sea, Surf and Sand: The Myth of the Beach*. Melbourne: Oxford University Press.

Elias, N. (1994) *The Civilising Process*. Oxford: Blackwell.

Evers, C. (2008) The Cronulla race riots: Safety maps on an Australian beach. *South Atlantic Quarterly* 107 (2), 411–429.

Fagan, R. and Webber, M. (1999) *Global Restructuring: The Australian Experience*. Melbourne: Oxford University Press.

Fiske, J., Hodge, B. and Turner, G. (1987) *Myths of Oz: Reading Australian Popular Culture*. Sydney: Allen and Unwin.

Fleming, A. (1969) *Brighton Beach Wollongong*. Wollongong, New South Wales: Illawarra Historical Society.

Forth, C.E. (2001) Moral contagion and the ill: The crisis of masculinity in the *fin-de-siècle* France. In A. Bashford and C. Hooker (eds) *Contagion: Historical and Cultural Studies* (pp. 61–73). London: Routledge.

Foucault, M. (1972) *The Archaeology of Knowledge and the Discourse on Language* (A.M. Sheridan Smith, trans.). New York: Pantheon.

Foucault, M. (1977) *Discipline and Punish: The Birth of the Prison* (A. Sheridan, trans.). London: Allen Lane.

Foucault, M. (1978) *The History of Sexuality. Volume I: An Introduction* (R. Hurley, trans.). London: Penguin Books.

Foucault, M. (1979) Governmentality. *Ideology and Consciousness* 6, 5–22.

Foucault, M. (1980) *Power/Knowledge: Selected Interviews and Other Writings, 1972–1977*. New York: Harvester Wheatsheaf.

Foucault, M. (1988) Technologies of the self. In L.H. Martin *et al.* (eds) *Technologies of the Self: A Seminar with Michel Foucault* (pp. 16–49). Amherst: University of Massachusetts Press.

Foucault, M. (1991) Governmentality. In G. Burchell, C. Gordon and P. Miller (eds) *The Foucault Effect: Studies in Governmentality with Two Lectures by an Interview with Michel Foucault* (pp. 87–104). Chicago: University of Chicago Press.

Fraser, L. (1994) *Losing it: America's Obsession with Weight and the Industry that Feeds on it*. New York: Dutton.

Frohlick, S. (2007) The negotiation of intimacy between tourist women and local men in a transnational town in Caribbean Costa Rica. *City and Society* 19 (1), 139–168.

Frohlick, S. (2008) 'I'm more sexy here': Erotic subjectivities of female tourists in the 'Sexual Paradise' of the Costa Rican Caribbean. In T.P. Uteng and T. Cresswell (eds) *Gendered Mobilities* (pp. 129–142). Aldershot: Ashgate.

Game, A. (1990) Nation and identity: Bondi. *New Formations* 11, 105–120.

Game, A. (1991) *Undoing the Social: Towards a Deconstructive Sociology*. Milton Keynes: Open University Press.

Garbarino, J. (1999) Lost boys. *Forum for Applied Research and Public Policy* 14 (4), 74–78.

Gibbs, A. and Warne, C. (1995) *Wollongong: A Pictorial History*. Alexandria, NSW: Kingsclear Books.

Gibson, M. (2001) Myths of Oz cultural studies: The Australian beach and English ordinariness. *Continuum: Journal of Media and Cultural Studies* 15 (3), 275–288.

Grosz, E. (1994) *Volatile Bodies: Towards a Corporeal Feminism*. St Lenoards: Allen and Unwin.

Hagan, J. (1997) Politics in the Illawarra. In J. Hagan and A. Wells (eds) *A History of Wollongong* (pp. 157–176). Wollongong: The University of Wollongong Press.

Hammond, R. (1913) Heliotherapy (of Rollier) as an adjunct in the treatment of bone disease. *The Journal of Bone and Joint Surgery* 2–11, 269–275.

Harper, M. (2007) *The Ways of the Bushwalker.* Sydney: UNSW Press.
Hartley, J. and Green, J. (2006) The public sphere of the beach. *European Journal of Cultural Studies* 9 (3), 341–361.
Harvey, D. (1989) *The Condition of Postmodernity: An Enquiry into the Origins of Cultural Change.* Cambridge: Blackwell.
Haug, J. (2005) *The History of the Bathing Suit*, accessed 6 Janaury 2011. http://www.victoriana.com/library/Beach/FashionableBathingSuits.htm
Heath, D. (2010) *Purifying Empire: Obscenity and the Politics of Moral Regulation in Britain, India and Australia.* Cambridge: Cambridge University Press.
Hemingway, J. (2006) Sexual learning and the seaside: Relocating the 'dirty weekend' and teenage girls' sexuality. *Sex Education* 6 (4), 429–433.
Hogan, M. (2008) *Labor Pains: Early Conference and Executive Reports of the Labor Party in New South Wales, 1906–1911*. Sydney: The Federation Press.
Horne, J. (2005) *The Pursuit of Wonder: How Australia's Landscape was Explored, Nature Discovered and Tourism Unleashed*. Victoria: Miegunyah Press.
Hubbard, P. (2000) Desire/disgust: Moral geographies of heterosexuality. *Progress in Human Geography* 24 (2), 191–217.
Hudson, B.J. (2000) The experience of waterfalls. *Australian Geographical Studies* 38 (1), 71–84.
Huntsman, L. (2001) *Sand in our Souls: The Beach in Australian History*. Victoria: Melbourne University Press.
Hutton, M. (1997) *Conservation Study for Belmore Basin Conservation Area, Wollongong, New South Wales*. NSW: Wollongong City Council.
Inglis, F. (2000) *The Delicious History of the Holiday*. London: Routledge.
Jackson, S. (1999) *Heterosexuality in Question*. London: Sage.
Jaggard, E. (1997) Chameleons in the surf. *Journal of Australian Studies* 21 (53), 183–191.
James, P. (1983) *Lifesaver*. Sydney: Lansdowne.
Jeans, D. (1990) Beach resort morphology in England and Australia: A review and extension. In P. Fabbri (ed.) *Recreational Uses of Coastal Areas* (pp. 277–285). Dordrecht: Kluwer.
Johnston, L. (2001) (Other) Bodies and tourism studies. *Annals of Tourism Research: A Social Science Journal* 28 (1), 180–201.
Johnston, L. (2005a) *Queering Tourism: Paradoxical Performances at Gay Pride Parades*. New York: Routledge.
Johnston, L. (2005b) Transformative tans: Gendered and raced bodies on beaches. *New Zealand Geographer* 61 (2), 110–116.
Johnston, L. and Longhurst, R. (2010) *Space, Place and Sex: Geographies of Sexualities*. Lanham: Rowman and Littlefield Publishers.
Kaplan, M. (1997) *Sexual Justice: Democratic Citizenship and the Politics of Desire.* New York: Routledge.
Kelly, D. (1997) Work and leisure, 1828–1997. In J. Hagan and A. Wells (eds) *A History of Wollongong* (pp. 177–188). Wollongong: The University of Wollongong Press.
Lash, S. and Urry, J. (1994) *Economies of Signs and Space*. London: Sage.
Lee, H. (1997a) Rocked in the cradle: The economy, 1828–1907. In J. Hagan and A. Wells (eds) *A History of Wollongong* (pp. 35–52). Wollongong: The University of Wollongong Press.
Lee, H. (1997b) A corporate presence: The economy, 1908–1945. In J. Hagan and A. Wells (eds) *A History of Wollongong* (pp. 53–70). Wollongong: The University of Wollongong Press.

Lewi, H. and Nichols, D. (eds) (2011) *Community: Building Modern Australia*. Randwick, Sydney: University of New South Wales Press.

Lewis, C. and Pile, S. (1996) Woman, body, space: Rio carnival and the politics of performance. *Gender, Place and Culture* 3 (1), 23–41.

Local Government Act (1906) Ordinance No. 52. Public Baths and Bathing.

Local Government Act (1919) Ordinance No. 52. Public Baths and Bathing.

Löfgren, O. (1999) *On Holiday: A History of Vacationing*. Berkeley: University of California Press.

Longhurst, R. (2001) *Bodies: Exploring Fluid Boundaries*. London: Routledge.

Lynch, R. and Veal, A. (1996) *Australian Leisure*. South Melbourne: Longman.

Macquarie, L. (1956) Lachlan Macquarie Governor of New South Wales. *Journals of His Tours in New South Wales and Van Diemen's Land 1810–1822*. Sydney Trustees of the Public Library of New South Wales.

Malam, L. (2004) Performing masculinity on the Thai beach scene. *Tourism Geographies* 6 (4), 455–471.

Massey, D. (2005) *For Space*. London: Thousand Oaks.

Matthews, G. (1994) *The Philosophy of Childhood*. Cambridge: Harvard University Press.

Maxwell, C. (1949) *Surf: Australians against the Sea*. Sydney: Angus and Robertson.

McClintock, A. (1993) *Imperial Leather: Race, Gender and Sexuality in the Colonial Context*. New York: Routledge.

McDermott, M. (2005) Changing visions of baths and bathers: Desegregating ocean baths in Wollongong, Kiama and Gerringong. *Sporting Traditions* 22 (1), 1–20.

McGloin, C. (2005) Surfing nation(s) – Surfing country(s). PhD thesis, University of Wollongong.

McGregor, C. (1994) The beach, the coast, the signifier, the Feral transcendence and pumpin' at Byron Bay. In D. Headon, J. Hooton and D. Horne (eds) *The Abundant Culture: Meaning and Significance in Everyday Australia* (pp. 51–60). St. Leonards, New South Wales: Allen and Unwin.

Menhennet, A. (1981) *The Romantic Movement*. London: Croom Helm.

Mercer, D. (1980) *In Pursuit of Leisure*. Victoria: Sorrett Publishing.

Michelson, P. (1993) *Speaking the Unspeakable: A Poetics of Obscenity*. Albany: State University of New York Press.

Middleton, R. and Figtree, A. (1963) *The History of the Growth of Surf Life-Saving Clubs on the Illawarra Coast of New South Wales*. Illawarra Branch, New South Wales: Surf Life Saving Association of Australia.

Mitchell, G. (1997) The garden of the Illawarra. In J. Hagan and A. Wells (eds) *A History of Wollongong* (pp. 143–156). Wollongong: The University of Wollongong Press.

Mooney, A. (2005) Keeping on the windy side of the law: The law of the beach. *Law Text Culture* 9 (1), 189–214.

Morgan, S. (2007) Wild oats or acorns? Social purity, sexual politics and the response of the late-Victorian church. *Journal of Religious History* 31 (20), 151–168.

Morris, M. (1992) On the beach. In L. Grossberg, C. Nelson and P. Triechler (eds) *Cultural Studies* (pp. 450–478). New York: Routledge.

Morris, M. (1998) *Too Soon, Too Late: History in Popular Culture*. Bloomington: Indiana University Press.

Moss, P. and Dyck, I. (1996) Inquiry into environment and body: Women, work and chronic illness. *Environment and Planning D: Society and Space* 14 (6), 737–753.

Munt, S. (1995) The lesbian flâneur. In D. Bell and G. Valentine (eds) *Mapping Desire: Geographies of Sexualities* (pp. 114–125). London: Routledge.

Nast, H. and Pile, S. (1998) *Places Through the Body*. London: Routledge.

New South Wales Government Gazette (1833) No. 50, February 13.

New South Wales Government Gazette (1838) No. 352, August 22.

New South Wales Legislative Council (1894) *Minutes and Proceedings*. New South Wales, Australia: Government Printer.

Orchard, K., Organ, M. and Walsh, J. (1994) *Illawarra – The Garden of New South Wales*. Wollongong City Art Gallery, 25 November 1994 [exhibition catalogue].

Organ, M.K. (1988) *The Illawarra Diary of Lady Jane Franklin*, 10–17 May 1839, accessed 3 March 2011. http://ro.uow.edu.au/asdpapers/34

Organ, N. and Speechley, C. (1997) Illawarra Aborigines. In J. Hagan and A. Wells (eds) *A History of Wollongong* (pp. 7–22). Wollongong: The University of Wollongong Press.

Pearson, K. (1979) *Surfing Subcultures of Australia and New Zealand*. Queensland: University of Queensland Press.

Perera, S. (2007) Aussie luck: The border politics of citizenship post Cronulla beach. *Australian Critical Race and Whiteness Studies e-journal* 3 (1), 1–16, accessed 3 March 2011. http://acrawsa.org.au/files/ejournalfiles/64SuvendriniPerera.pdf

Perera, S. (2009) *Australia and the Insular Imagination: Beaches, Border, and Boats*. New York: Palgrave Macmillan.

Phillips, J. (2002) The beach boy of Barbados: The postcolonial entrepreneur. In S. Thorbek and B. Pattanaik (eds) *Transnational Prostitution: Changing Patterns in a Global Context* (pp. 42–56). New York: Zed Books.

Phillips, L. and Jørgensen, M. (2002) *Discourse Analysis as Theory and Method*. Thousand Oaks: Sage.

Phillips, R. (1997) *Mapping Men and Empire: A Geography of Adventure*. London: Routledge.

Phillips, R. (2006) *Sex, Politics and Empire: A Postcolonial Geography*. Manchester: Manchester University Press.

Probyn, E. (2000) *Carnal Appetites: Food Sex Identities*. New York: Routledge.

Probyn, E. (2003) The spatial imperative of subjectivity. In K. Anderson, M. Domosh, S. Pile and N. Thrift (eds) *Handbook of Cultural Geography* (pp. 290–299). London: Sage.

Ramsland, J. (2000) They ride the surf like gods: Sydney-side beach culture, life-saving and eugenics, 1902–1940. In M. Crotty, J. Germov and G. Rodwell (eds) *A Race for a Place: Eugenics, Darwinism and Social Thought and Practice in Australia* (pp. 263–273). New Castle: Faculty of Arts and Social Science, University of Newcastle.

Rickard, J. (1998) Lovable larrikins and awful ockers. *Journal of Australian Studies* 22 (56), 78–85.

Rodwell, G. (1999) The sense of victorious struggle: The eugenic dynamic in Australian popular surf-culture, 1900–50. *Journal of Australian Studies* 23 (62), 56–63.

Rose, G. (1993) *Feminism and Geography: The Limits of Geographical Knowledge*. Cambridge: Polity Press.

Rose, N. (1996) Governing 'advanced' liberal democracies. In A. Barry, T. Osborne and N. Rose (eds) *Foucault and Political Reason: Liberalism, Neo-Liberalism and Rationalities of Government* (pp. 37–64). Chicago, IL: University of Chicago Press.

Rosenthal, M. (1986) *The Character Factory*. London: Collins.

Russell, P.A. (2010) *Savage or Civilised? Manners in Colonial Australia*. Sydney: University of New South Wales Press.

Ryan, J.R. (1997) *Picturing Empire: Photography and the Visualization of the British Empire*. Chicago: University of Chicago Press.

Said, E. (1993) *Culture and Imperialism*. London: Chatto and Windus.

Schama, S. (1995) *Landscape and Memory*. New York: Alfred A. Knopf.

Scott, S. (2010) How to look good (nearly) naked: The performative regulations of the swimmer's body. *Body and Society* 16 (2), 143–168.

Segal, L. (1990) *Slow Motion: Changing Masculinities, Changing Men*. New Brunswick, NJ: Rutgers University Press.

Sheldon, P. (1997) Local Government to 1947. In J. Hagan and A. Wells (eds) *A History of Wollongong* (pp. 101–114). Wollongong: The University of Wollongong Press.

Shields, R. (1991) *Places on the Margin: Alternative Geographies of Modernity*. London: Routledge.

Smith, D. (2000) *Moral Geographies: Ethics in a World of Difference*. Edinburgh: Edinburgh University Press.

Smith, N. (1992) Geography, difference and the politics of scale. In D. Doherty, E. Graham and M. Malek (eds) *Postmodernism and the Social Sciences* (pp. 57–59). London: Macmillan.

Soloway, R.A. (1995) *Demography and Degeneration: Eugenics and the Declining Birthrate in Twentieth-Century Britain*. Chapel Hill, NC: The University of North Carolina Press.

Summers, A. (1975) *Damned Whores and God's Police*. Ringwood, Victoria: Penguin.

Turner, B. (1984) *The Body and Society: Explorations in Social Theory*. Oxford: Blackwell.

Urry, J. (1990) *The Tourist Gaze: Leisure and Travel in Contemporary Societies*. London: Sage Publications.

Urry, J. (2000) *Sociology Beyond Societies: Mobilities for the Twenty-First Century*. London: Routledge.

Valentine, G. (1989) The geography of women's fear. *Area* 21 (4), 385–390.

Valentine, G. (1996) Children should be seen and not heard: The production and transgression of adult's public space. *Urban Geography* 17 (3), 205–220.

Valentine, G. (2001) *Social Geographies: Space and Society*. New York: Prentice-Hall.

Waitt, G. and Markwell, K. (2006) *Gay Tourism: Culture and Context*. New York: Haworth Press.

Waitt, G. and Gorman-Murray, A. (2008) Camp in the country: Renegotiating sexuality and gender through a rural lesbian and gay festival. *Journal of Tourism and Cultural Change* 6 (3), 185–207.

Waitt, G. and Stapel, C. (2011) 'Fornicating on floats?' The cultural politics of the Sydney Mardi Gras Parade beyond the metropolis. *Leisure Studies* 30 (2), 197–216.

Walton, J. (1983) *The English Seaside Resort: A Social History 1750–1914*. New York: Leicester University Press.

Ward, C. and Hardy, D. (1986) *Goodnight Campers!: The History of the British Holiday Camp*. London: Mansell.

Waterhouse, R. (1995) *Private Pleasures, Public Leisure: A History of Australian Popular Culture Since 1788*. Melbourne: Longman.

Wells, L. (1982) *Sunny Memories: Australians at the Seaside*. Victoria: Greenhouse Publications.

White, C. (2003) Picnicking, surf-bathing and middle-class morality on the beach in the eastern suburbs of Sydney, 1811–1912. *Journal of Australian Studies* 27 (80), 101–110.

White, C. (2006) Promenading and picnicking: The performance of middle-class masculinity in nineteenth-century Sydney. *Journal of Australian Studies* 30 (89), 27–40.

White, C. (2007) Save us from the womanly man: The transformation of the body on the beach in Sydney, 1810 to 1910. *Men and Masculinities* 10 (1), 22–38.

Williams, R. (1982) *The Sociology of Culture*. New York: Schocken Books.

Woollacott, A. (2008) London, New York and Hollywood: Three 'Australians' on the world stage. In R. Dixon and V. Kelly (eds) *Impact of the Modern: Vernacular Modernities in Australia 1870s–1960s* (pp. 185–201). Sydney: Sydney University Press.
Wyndham, D. (2003) *Eugenics in Australia: Striving for National Fitness.* London: The Galton Institute.
Young, I.M. (1990) *Throwing Like a Girl and Other Essays in Feminist Thought.* Bloomington and Indianapolis: Indiana University Press.

Guidebooks

Beautiful Illawarra: The Illawarra or South Coast Tourist Guide (1899) Sydney: The South Coast Tourist Union, Andrews and Cook, Pitt Street.
The Beautiful Illawarra District, Wollongong. The Ideal Seaside and Tourists' Resort (1910) Printed at the South Coast Times Office, Crown Street, Wollongong, December 1910 (no pagination).
By Train in Daylight through the Beautiful Illawarra (c. 1930) New South Wales: Commissioner for Railways (no pagination).
The Illawarra or South Coast Tourist Guide (1903) The South Coast Tourist Union, Sydney: John Sands, Printer, George Street.
The Illawarra Tourist Guide: Wollongong Citizens Association (c. 1920s) Crown Street, Wollongong: Hamey Studios, Epworth Press.
Picturesque Illawarra: The Garden of New South Wales (1931) Issued by the Illawarra Tourist and Publicity Bureau and the Illawarra Motor and Tourist Services, prepared by Michael P. Hennessy.
Wilson's Rail, Road and Sea Guide to the South Coast and Southern Highlands, New South Wales (1929) 3 Spring Street, Sydney: Wilson's Publishing Company.

Magazines

The Australian Women's Weekly (1933a) MORE DARING surf suits than EVER BEFORE by Nell Murray. 19 August, p. 2.
The Australian Women's Weekly (1933b) Seaside holiday: Sunburn and laughter ... and a dipping sea. 25 November, p. 1.
The Australian Women's Weekly (1933c) Seaside CHILDREN and their City COUSINS. 9 December, p. 12.
The Australian Women's Weekly (1933d) The law and Mrs. Grundy. 21 October, p. 4.
The Australian Women's Weekly (1933e) So they say. D.A. Jurns. 18 November, p. 17.
The Australian Women's Weekly (1933f) So they say. C. Halbert. 18 November, p. 17.
The Australian Women's Weekly (1933g) So they say. Miss J. Gilmore. 18 November, p. 17.
The Australian Women's Weekly (1933h) So they say. Miss S. Hyde. 14 October, p. 14.
The Australian Women's Weekly (1933i) Bare skin *Craze Sweeps* England. 30 September, p. 2.
The Australian Women's Weekly (1934a) So they say. Mrs. T. Mitchell. 15 December, p. 21.
The Australian Women's Weekly (1934b) Will men be allowed to wear trunks on beaches? 14 July, p. 4.
The Australian Women's Weekly (1934c) So they say. Mrs. E.G. Woodger. 28 July, p. 13.
The Australian Women's Weekly (1934d) So they say. Kay Hearfield. 28 July, p. 13.
The Australian Women's Weekly (1934e) So they say. Mrs. G.K. Yorkston. 28 July, p. 13.
The Australian Women's Weekly (1934f) So they say. Mrs. W. Foster. 28 July, p. 13.

The Australian Women's Weekly (1934g) The body beautiful, sun-tanned in the smart, shiny way! 22 September, pp. 37–38.
The Australian Women's Weekly (1934h) So they say. Miss Kathleen Court. 25 August, p. 22.
The Australian Women's Weekly (1935a) So they say. M. White. 5 January, p. 19.
The Australian Women's Weekly (1935b) The new Kava-Knits ... exclusively Jantzen! 28 September, p. 18.
The Australian Women's Weekly (1935c) The body beautiful, fear no more the heat o' the sun ... follow these golden rules for sunbaking! 28 September, p. 51.
The Australian Women's Weekly (1936a) Your Jantzen won't do a February fade out. 17 October, p. 48.
The Australian Women's Weekly (1936b) The body beautiful. When you go down to the sea, don't overcook yourself, fair maid, in a hurried, futile effort to acquire bronze loveliness. 12 December, p. 75.
The Australian Women's Weekly (1937a) Subtle flattery by Jantzen. 20 November, p. 36.
The Australian Women's Weekly (1937b) Outdoor loveliness for you! How to acquire a nice tan with no freckles or burns to spoil the effect of delicate party frocks. 30 October, p. 63.
The Australian Women's Weekly (1938a) Skip the flattery, darling – My Jantzen takes care of that. 22 October, p. 64.
The Australian Women's Weekly (1938b) Does sunbaking pay? Holidaymakers should take the sun in easy doses. 31 December, p. 4.
The Australian Women's Weekly (1940) If you get badly SUNBURNT. 20 January, p. 45.

Newspapers

Note: All *The Australian*; *Illawarra Mercury*; *Kiama Independent;* and *South Coast Times* newspapers in this list have no pagination.

The Australian (1838) Wollongong. J.T. Wilson, Sydney Auctioneer. 23 October.
Daily Telegraph (1908) Surf bathing. Its past, present and future. 13 February, p. 5.
Illawarra Mercury (1856) I'm going to bathe! Where? In Johnson's new bathing machine. The first ever seen in the colony!! 29 December.
Illawarra Mercury (1857) A bath! A bath!! Now for a bath!!! 5 January.
Illawarra Mercury (1858) A disgraceful practice. 11 March.
Illawarra Mercury (1869) Bathing place. 2 February.
Illawarra Mercury (1871) The new bathing place. 13 January.
Illawarra Mercury (1888) Hours for bathing. 11 December.
Illawarra Mercury (1902a) The searchlight. 8 January.
Illawarra Mercury (1902b) Kellerman Comedy Co. 18 January.
Illawarra Mercury (1907) Bathing costumes. 15 October.
Illawarra Mercury (1908a) Bathing ordinances. Now in force. 25 February.
Illawarra Mercury (1908b) Mixed surf-bathing (To the Editor). 14 January.
Illawarra Mercury (1909) Sun bathing. 5 March.
Illawarra Mercury (1911) Do tourists help Wollongong. 21 February.
Illawarra Mercury (1915) Bulli Shire Council. Correspondence. 19 November.
Illawarra Mercury (1917) Mixed bathing. At The Wollongong Swimming Baths (To the Editor). 5 January.
Illawarra Mercury (1928) Wollongong Continental Baths. Official opening. 30 March.
Illawarra Mercury (1929a) New South Wales Government Railways. Excursion to Sydney and intermediate stations. 25 October.

Illawarra Mercury (1929b) New Prince's Highway Guide for tourists. 27 December.
Illawarra Mercury (1929c) Bulli Shire Council. Correspondence. 29 November.
Illawarra Mercury (1929d) The Shellharbour Council. Monthly meeting. 20 December.
Illawarra Mercury (1929e) Wollongong Council. Park Committee's Report. 6 December.
Illawarra Mercury (1929f) Tourist Association. Wollongong Branch. 20 December.
Illawarra Mercury (1929g) North Wollongong Beach and the fence. 20 December.
Illawarra Mercury (1929h) Bulli Shire Council Regular Meeting. Correspondence. 4 October.
Illawarra Mercury (1929i) Surf life saving. 4 October.
Illawarra Mercury (1929j) Timely topics. (by Omar). 27 December.
Illawarra Mercury (1929k) Wollongong's Continental Baths. 11 January.
Illawarra Mercury (1929l) Special display of bathing and beach attire at Lances Ltd. 4 October.
Illawarra Mercury (1929m) Have you answered the call? Lances Ltd. Wollongong. 1 November.
Illawarra Mercury (1930a) Illawarra Surf Life Life (sic) Asscn. Delegate Meeting. 21 January.
Illawarra Mercury (1930b) Tourist Association. Correspondence. 7 February.
Illawarra Mercury (1930c) The Surf. Austinmer Surf News (By Linesman). 7 February.
Illawarra Mercury (1930d) Thirroul Surf Club. Queen Competition. 17 January.
Illawarra Mercury (1931a) Tourist Association. Wollongong Annual Meeting. 18 September.
Illawarra Mercury (1931b) The surf. North Beach (by "Sunshine"). 4 December.
Illawarra Mercury (1931c) Parks and Baths Committee Report. 30 January.
Illawarra Mercury (1931d) Bulli Shire Council. Fortnightly Meeting. Correspondence. 20 February.
Illawarra Mercury (1931e) The surf. North Beach News by Sunshine. 23 October.
Illawarra Mercury (1931f) The surf. Illawarra Branch S.L.S. Assn (By Beltman). 6 February.
Illawarra Mercury (1932) The ladies' baths at Wollongong. To the Editor. 15 January.
Illawarra Mercury (1935a) Attracting tourists. Interesting Address. 6 September.
Illawarra Mercury (1935b) The tourist traffic. 11 January.
Illawarra Mercury (1935c) Tourist matters. Meeting at Wollongong. 21 June.
Illawarra Mercury (1935d) Bulli Council. Fortnightly meeting. 18 January.
Illawarra Mercury (1935e) The surf. Wollongong S.L.S. Club Annual Meeting. 18 October.
Illawarra Mercury (1935f) The surf. Illawarra Branch S.L.S.A. 18 October.
Illawarra Mercury (1935g) The surf. Illawarra Branch Surf Life Saving Association (by P.O.). 5 April.
Illawarra Mercury (1935h) The surf. Port Kembla Club Notes (by "Sunbaker"). 20 December.
Illawarra Mercury (1935i) The surf. Surf Queen. 8 February.
Illawarra Mercury (1935j) The surf. Notes from North Beach (by "Farmer"). 27 September.
Illawarra Mercury (1935k) The surf. Notes from North Beach (by "Farmer"). 18 October.
Illawarra Mercury (1936a) Coast trip. A large contingent. Tablelands to Wollongong. 7 February.
Illawarra Mercury (1936b) Wollongong Council. Fortnightly meeting. 13 November.
Illawarra Mercury (1936c) New surf pavilion officially opened. 9 October.
Illawarra Mercury (1936d) The surf. Notes from W.S.L.S. Club. 3 April.
Illawarra Mercury (1936e) The surf. Corrimal Club (by Publicity Officer). 3 April.
Illawarra Mercury (1936f) The surf. Port Kembla (by "Sunbaker"). 31 January.
Illawarra Mercury (1937a) Cruising train visits South Coast. 10 September.
Illawarra Mercury (1937b) More railway excursions. 15 October.
Illawarra Mercury (1937c) Surf notes. Wollongong Club (by Kube). 22 October.

Illawarra Mercury (1937d) Surf notes. Austinmer Surf News (by Surf Ski). 24 December.
Illawarra Mercury (1937e) Surf notes. Notes from North Beach (by Surfo). 8 January.
Illawarra Mercury (1937f) Bulli Shire Council. Fortnightly meeting. Correspondence. 26 February.
Illawarra Mercury (1937g) Grand Hotel. 3 December.
Illawarra Mercury (1937h) Surf notes. Corrimal Club (by Shark). 8 October.
Illawarra Mercury (1937i) Surf notes. Wollongong Surf Club (by "Jack"). 5 March.
Illawarra Mercury (1937j) Surf notes. Illawarra Branch S.L.S.A. (by E.S. Lynch). 3 December.
Illawarra Mercury (1937k) Surf notes. Corrimal Club (by Shark). 22 October.
Illawarra Mercury (1937l) Surf notes. Austinmer Club (by "Surf Ski"). 10 December.
Illawarra Mercury (1937m) Surf notes. Illawarra Branch S.L.S.A. Wollongong Carnival (by E. Lynch). 5 March.
Illawarra Mercury (1937n) Carefree holidays ahead! LANCES say "You're your best in a JANTZEN". 10 December.
Illawarra Mercury (1937o) Bathing in trunks. Austinmer Seeks Prosecution. 12 February.
Illawarra Mercury (1938a) Surf notes. Austinmer Club (by "Surf Ski"). 7 January.
Illawarra Mercury (1938b) North Illawarra Council. Correspondence. 18 February.
Illawarra Mercury (1938c) Surf news. Illawarra Branch Notes (by A.J.T.). 16 December.
Illawarra Mercury (1938d) Surf news. Illawarra S.L.S. Assn. Annual Report. 21 October.
Illawarra Mercury (1938e) Surf news. Bulli Club (by Publicity Officer). 16 December.
Illawarra Mercury (1938f) Surf news. Austinmer Club (by "Surf Ski"). 16 December.
Illawarra Mercury (1938g) Surf notes. Corrimal Club (by Shark). 7 January.
Illawarra Mercury (1938h) Surf news. Austinmer Club (by Publicity Officer). 14 October.
Illawarra Mercury (1938i) Bulli Shire Council. Fortnightly meeting. Reports by Beach Inspector. 11 February.
Illawarra Mercury (1939a) Surf news. Bulli Club (by Breaker). 6 January.
Illawarra Mercury (1939b) Surf news. Illawarra Branch Notes (by A.J.T.). 3 November.
Illawarra Mercury (1939c) Surf news. Corrimal Club (by Rodger). 3 November.
Illawarra Mercury (1939d) Surf news. Wollongong Club (by Ern Lynch). 3 November.
Illawarra Mercury (1939e) Surf news. Notes from North Beach (by Dumper). 3 November.
Illawarra Mercury (1939f) Surf news. Woonona Club. Tribute to late club mate. 3 November.
Illawarra Mercury (1939g) LANCE'S the store for Jantzens. 27 October.
Kiama Independent (1888) Local news. 23 October.
Kiama Independent (1890) The Kiama swimming baths. 4 November.
Kiama Independent (1895) The penalty for bathing on beaches. 11 April.
Kiama Independent (1908) Kiama Council. Petition. 21 March.
South Coast Times (1902) Swimming carnival. 22 February.
South Coast Times (1908a) Wollongong Council. 25 January.
South Coast Times (1908b) Bathing regulations. To the Editor, South Coast Times: Lady Bather's Brother. 25 January.
South Coast Times (1908c) Bathing regulations. To the Editor, South Coast Times: Visitor. 25 January.
South Coast Times (1914a) Surfing ceremony. 16 October.
South Coast Times (1914b) Motor cars for hire. 16 October.
South Coast Times (1928) Official opening of the Continental Baths. 30 March.
South Coast Times (1936) New £6000 surf pavilion foundation stone laid. 6 March.
The Sun (1931) The thing that spells freedom in the true sense of the word. 15 November, p. 23.

The Sun (1932) Mystery hikes. 8 August, p. 8.
Sydney Mail (1888) Opening of the line to Wollongong and Kiama. 13 October, p. 770.
Sydney Mail (1906a) The summer seaside carnival. 7 March, p. 604.
Sydney Mail (1906b) A view on the South Coast. 7 March, p. 620.
Sydney Mail (1906c) In the breakers. 7 March, p. 619.
Sydney Morning Herald (1842) Illawarra. 3 January (no pagination).
Sydney Morning Herald (1907a) Surf bathing To the Editor of the Herald: G. Norton Russel. 1 February, p. 5.
Sydney Morning Herald (1907b) Death of surf bathing. 15 October, p. 5.
Sydney Morning Herald (1907c) Surf bathing at Manly by W. Tonge and A.W. Relph. 7 September, p. 6.
Sydney Morning Herald (1907d) Surf bathing To the Editor of the Herald: Daily dipper. 1 February, p. 5.
Sydney Morning Herald (1908) Heroes of the surf. Life-saving methods by A.W. Relph. 26 September, p. 7.
Sydney Morning Herald (1909) Surf-bathing by A.W. Relph. 11 September, p. 5.

Unfiled records

Council records

Council minutes books reporting on the Parks and Baths Committee and Progress Association Committee: 1886–1940 (Bulli Shire, Central Illawarra Municipality/ Shire, North Illawarra Municipality and Wollongong Municipality).

Surf Life Saving Association of the Illawarra branch records

Annual Reports of the Illawarra Branch of the SLSA (1932/33 – 43/44).
Club histories from the following surf clubs: Bellambi, Coal Cliff, Corrimal, Bulli, Thirroul, Wollongong and Helensburgh.
Unfiled surf club material 1908–1940.

New South Wales Government Railways records

Annual Report on Traffic Branch Operations, 1940
New South Wales Government Railways passenger fares and coaching rates 1920, 1922
Special Train Notices, 1920, 1930
Unfiled material for the South Coast Line: 1900–1940s, held at the State Records Office, Penrith.
Weekly Special Train Notices, 1930

Bank of New South Wales records

For the period 1900–1940 (available) bank manager reports were examined for every year for Wollongong, Port Kembla and Kiama branches and Corrimal, Woonona and Crown Street, West Wollongong branch agencies.
Inspector's Report Metropolitan Division New South Wales: 1909–1931.
Inspector's Report, Bank of New South Wales Central Division: 1932–1940.